13032

WELLINGTON COLLEGE, BELFAST. LIBRARY

Date Due	Date Due	Date Due

AN INTRODUCTION TO
SEMICONDUCTORS

An Introduction to Semiconductors

K. J. CLOSE, B.SC., PH.D., F.INST.P.
Principal Lecturer in Physics,
The Polytechnic of Central London

J. YARWOOD, M.SC., F.INST.P.
Head of Department of Mathematics and Physics,
The Polytechnic of Central London

HEINEMANN EDUCATIONAL BOOKS
LONDON AND EDINBURGH

Heinemann Educational Books Ltd
LONDON EDINBURGH MELBOURNE TORONTO
AUCKLAND JOHANNESBURG SINGAPORE
HONG KONG IBADAN NAIROBI NEW DELHI

ISBN 0 435 68080 3 (*cased edition*)
     0 435 68080 1 (*limp edition*)

© K. J. Close and J. Yarwood 1971

First published 1971

Published by Heinemann Educational Books Ltd
48 Charles Street, London W1X 8AH

Printed in Great Britain at the Pitman Press, Bath

Preface

The junction diode, the various types of transistor, and other semiconductor devices have largely replaced the thermionic vacuum tubes, and especially the small power tubes, in the majority of applications in electronics.

It is relatively easy to give a simple account of the radio valves such as the triode and the pentode together with their easier circuit applications which can be understood by the beginner. It is not so easy to do this for the semiconductor devices. Yet it is imperative, even at the level of elementary physics in school, that this should be done, because otherwise the student's appreciation of modern electronics will be hopelessly out of date.

The syllabuses in physics of the various 'O' Level, 'A' Level, Ordinary National Certificate, and City and Guilds examinations are therefore desirably beginning to include elementary aspects of semiconductor physics. This tendency will inevitably increase in the future.

It is considered important that the student should not simply accept categorical statements about these devices, but gain a reasonably fundamental understanding of their operation which is based as far as is practicable on other aspects of physics in the course. The study should therefore be based on fundamental principles not only because of the educational value of such a procedure but also because this is the only way to ensure that the student will be able to understand the transistor device of a decade hence, likely to be significantly different from the ubiquitous junction transistor of today. The course should also include relevant simple calculations, exercises and simple practical experiments, yet must avoid the complexities of sophisticated topics such as Fermi-Dirac statistics and the quantum mechanics of conduction processes.

It is with these aims in view that this book is presented. It has been written with a paramount idea in mind: that a student at A-Level standard should be able to understand it. It is not expected, however, that all the material included should be dealt with in a school course; some of it overlaps into the first year of science or technology at College or University.

It is hoped that this text will assist beginners in appreciating the fascination and utility of these devices which have changed electronics so greatly and will undoubtedly continue to do so in the future.

1971

K.J.C.
J.Y.

Contents

	Page
PREFACE	v

1. The structure of matter and the solid state — 1

The states of matter; crystalline structure; motion of electrons; the kinetic theory of gases; atomic structure; the periodic table; electron gas theory; electrical properties of metals; semiconductors; intrinsic and extrinsic semiconductors; carrier mobility and current density; energy levels and energy bands; insulators; energy gap determination; the Hall effect; identification of majority carriers.

2. The manufacture of semiconductor devices — 53

Zone refining; growing large single crystals; forming a p-n junction; rate-growing; alloy junctions; junctions produced by diffusion.

3. Semiconductor diodes and rectification — 62

Power supply; the p-n junction diode and its electrical behaviour; half- and full-wave rectification; voltage doubling and quadrupling; smoothing filters; Zener diodes; voltage regulation; further applications of Zener diodes; shunting a current meter to provide a non-linear scale.

4. The junction transistor — 97

Structure and behaviour; common-emitter and common-base connection; hybrid parameters; amplification; the load-line; equivalent circuits; the decibel; leakage current and thermal runaway; biasing arrangements; a two-stage amplifier; feedback; a constant-current source; stabilized power supplies; transistor oscillators; the multi-vibrator; a simple transistor tester.

Contents

                                                                            Page

5. Other Semiconductor Devices   140

The unijunction transistor; a relaxation oscillator; a staircase generator; the silicon controlled rectifier; a.c. phase control; a bistable circuit; a lamp dimming circuit; field effect transistors; semiconductor particle detectors; photoelectric effects in semiconductors; thermistors.

ANSWERS   177

INDEX   179

1 The structure of matter and the solid state

1.1 Introduction

The study of semiconductors forms a part of the wider subject known as solid-state physics. The concern is with the passage of electric current through solids as opposed to liquids and gases.

The fundamental electric charge is that of the electron, a particle of mass m_e approximately 9×10^{-31} kg and charge e equal to 1.6×10^{-19} coulomb. The number of electrons in one coulomb of electric charge is thus $1/(1.6 \times 10^{-19})$, equal to 6.3×10^{18} approximately. As the ampere, the unit of electric current, corresponds to the passage of one coulomb of charge per second, so one ampere represents a transfer of 6.3×10^{18} electrons per second.

In this text the electron will be treated as a particle, even though on many occasions in solid-state physics, the location and behaviour of electrons have to be inferred by regarding the moving electron as a wave packet, leading to the methods of wave mechanics originated by the French physicist Louis de Broglie in 1924.

Once the existence of the electron was established finally in 1897 by Sir J. J. Thomson, much attention was devoted to explain the electrical properties of materials by the motion of electrons. Of particular interest were the metals, known to be very good conductors and therefore assumed to contain many 'free' electrons available for conduction, where by 'free' is meant that these electrons are bound in the metal by such insignificant forces that they are set in motion even by the smallest applied electric fields. Insulators were assumed to have no or very few free electrons available within their structure, whilst semiconductors, with electrical conductivities intermediate between those of metals and insulators, were represented by some intermediate state of electron containment.

In the case of metals, each atom may contribute one, two or even three free electrons to serve as current carriers. If each atom contributes only one free electron, there will be as many as 10^{29} free electrons per cubic metre of the metal. These electrons are able to move through the metal on the application of an electric field, but their motion will be restricted because they collide with the atoms of the metal. Their motion is similar to that of molecules in a gas, so similar, in fact, that some of the concepts and equations from the kinetic theory of gases were used to develop a theory of metallic conduction. This *free electron gas theory*, as it was called, was partly successful when applied to metals but failed to explain the electrical behaviour of semiconductors.

An understanding of electrical conduction in solids thus demands a study of the electron as a charge carrier, the part it plays in atomic structure, the nature of interatomic forces and the arrangements of atoms in materials. In addition, certain aspects of the kinetic theory, originally applied to gases, are seen to be relevant.

1.2 The Three States of Matter

A concentration of the atoms of an element can be arranged to be in a solid, liquid or gaseous state simply by controlling the environment in which the element is situated. In a solid element, the atoms are held in position by very strong mutual attractive forces which are electrical in origin. Their positions relative to one another in the solid at a given temperature will be in equilibrium when the system is in its lowest energy state. Work must be done to distort this equilibrium arrangement. This work is needed to alter the spacing between atoms. Because the displaced atoms will tend to return to their equilibrium positions, the solid is elastic. For example, the forces required to stretch a metallic rod show that the attractive forces between its atoms are considerable. Again, if any attempt is made to compress a solid, an elastic behaviour indicates the very strong repulsive forces between atoms which are brought into play when the spacings between atoms at equilibrium are decreased by the compression.

The atoms of a solid vibrate about their equilibrium positions with an amplitude which increases with temperature. An increase in temperature at first results in thermal expansion but, at the melting point, the rigid structure breaks down, the latent heat of fusion representing the energy needed to convert the solid into a liquid.

In the liquid state, which is the most complex of the three states of matter to study, the atoms move in random fashion, yet are still subjected to mutual attractions which retain the constant volume of the sample. That the motion of the atoms is very rapid is demonstrated by the ease with which diffusion occurs in liquids; diffusion also takes place in solids but at an almost imperceptible rate because of the constraint imposed by neighbouring atoms held in the rigid structure. Further evidence of the mutual attractive forces between atoms in a liquid is provided by the existence of surface tension forces.

At any temperature, whether the element is in a solid, liquid or gaseous state, the energies of the atoms are distributed about some average energy, which increases with the temperature. At any instant of time, some atoms will have energies much greater and some much less than this average value. Any fast-moving atoms near the surface of a solid or liquid may escape into the space above; solids and liquids therefore exert a vapour pressure, though that of liquids is normally much larger than that of solids. On evaporation from a liquid, only the most energetic atoms can escape, so that the average energy of all those atoms remaining behind must fall. The liquid temperature will therefore fall, and hence a liquid cools on evaporation.

Further increase of temperature of the element beyond its melting point will increase the rate of evaporation. At the boiling point, which increases with the external pressure, the liquid is converted into vapour; this process requires an amount of heat called the latent heat of vaporization of the liquid. The forces of attraction between the atoms of the vapour are now very small compared with those which prevailed in the liquid state, and the vapour will now fill any container in which it is placed. The atoms can now travel significant distances between collisions, and these distances increase as the pressure is reduced. The motions of the atoms and their behaviour in the gaseous state is more readily understood than in the solid or liquid state.

In this account, atoms of an element have been discussed: similar statements apply to the molecules of a compound. The metallic elements have monatomic molecules, but an element such as hydrogen, oxygen or nitrogen will normally exist in the form of diatomic molecules.

1.3 Crystalline Structure

Many common substances are crystalline. Familiar examples are common salt (NaCl) and sugar. Each particle is a crystal having a specific geometrical form. Among the large crystals occurring in nature are quartz, diamond and mica. Metals are crystalline. In general, they are made up of a very large number of tiny crystals so small that a microscope is needed to discern them, though large crystals of metals can be grown by special techniques. X-rays can be used to examine the structure both of large single crystals and even of microcrystals which are not easily identified, even though discernable, under a microscope.

In a crystal, the atoms are arranged in a very orderly geometrical pattern known as the *crystal lattice*. All pure crystals of the same element have the same lattice structure. Solids in which there is no such definite arrangement are said to be *amorphous*, an example being glass.

The separation between neighbouring planes of atoms often differs in different directions in the crystal; thus, a different value for the elastic modulus can be measured in each of these directions. A specimen which exhibits different properties in different directions is called *anisotropic*. However, whereas single crystals are highly anisotropic, a specimen made up of millions of tiny crystals orientated at random, i.e. a polycrystalline specimen, is, in general, isotropic, a familiar example being an ordinary piece of copper, aluminium or iron.

There are seven basic patterns of crystal of which only one, the cubic, is considered here. Many materials have a cubic crystal structure. Two examples are common salt or sodium chloride (NaCl) and potassium chloride in the form of sylvine (KCl).

The cubic crystal of NaCl is formed from an enormous number in adjacent cubic elements each of which is of the form shown of Figure 1.1(a). At each corner of the elemental cube there is an atom and these atoms are arranged to be alternately of sodium (Na) and chlorine (Cl). In NaCl, the valence bond is between the electropositive sodium and the electronegative chlorine. The sodium atom gives up its outermost valence electron to the chlorine atom so that the first is a positive ion of sodium and the second a negative ion of chlorine (Figure 1.1(b)).

The dimensions of such an elemental cube are very small. In the case of NaCl the side of the cube is only 0.282 nm (1 nm = 10^{-9}m = 10 Ångstrom unit). Even in a tiny microcrystal there are several million such elemental cubes, and these cubes are arranged in the crystal as shown in Figure 1.1(c), forming a crystal lattice in which it is noted that each atom (or rather ion) is at a point forming the corner of eight adjacent cubes.

The crystal lattice in this form is known as *simple cubic*. *Face-centred cubic* (fcc) and *body-centred cubic* (bcc) structures (Figure 1.1(d) and (e) respectively) are also frequently encountered. Many metals have crystals of one of these forms, for example, tungsten (bcc), molybdenum (bcc) and platinum (fcc).

Imperfections in a crystal influence greatly its physical and especially its electrical properties. A synthetic gemstone has only a fraction of the monetary value of a natural gemstone and yet it is a more perfect crystal. The attractive colour and lustre of a natural stone results from the presence of impurity atoms in the lattice.

If a foreign atom occupies a lattice site normally occupied by the host atom, it is called a *substitutional impurity*. On the other hand, if the foreign atom is lodged between the lattice planes of the host atoms it is said to be an *interstitial impurity*.

To change the electrical properties of a semiconducting crystal in a controlled fashion, foreign atoms of a selected element are introduced into substitutional sites in the host lattice. The techniques of introducing such foreign atoms into the pure crystal are described in Chapter 2.

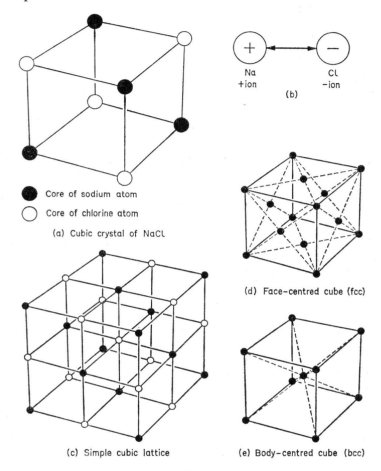

Figure 1.1. Cubic crystal structures

1.4 The Motion of Free Electrons in an Electric Field

Because, by definition, the electric field strength at a point is the force on unit positive electric charge placed at that point, the force **F** (the bold print denotes a vector quantity) acting on a positive charge q in an electric field of strength **E** is given by

$$\mathbf{F} = q\mathbf{E} \tag{1.1}$$

where the force acts in the direction of the field. The force is in newton if q is in coulomb and E is in volt per metre.

If V is the potential at this point, the field strength $\mathbf{E}_x$ is related to the potential gradient by

$$\mathbf{E}_x = \frac{-\mathrm{d}V}{\mathrm{d}x} \tag{1.2}$$

where $\mathbf{E}_x$ is the component of the field strength in the x direction. Therefore

$$\mathbf{F}_x = -q\frac{\mathrm{d}V}{\mathrm{d}x} \tag{1.3}$$

For an electron with a negative charge of magnitude e the force is

$$\mathbf{F}_x = e\frac{\mathrm{d}V}{\mathrm{d}x} \tag{1.4}$$

and is in the direction of V increasing. Under the action of this force, a free electron moves and gains energy. The work done by the force in moving its from point 1 at potential V_1 to point 2 at potential V_2 is seen from equation (1.4) to be

$$\int_1^2 F_x \mathrm{d}x = \int_{V_1}^{V_2} e\, \mathrm{d}V = eV_2 - eV_1 \tag{1.5}$$

This work done will give kinetic energy to the electron. The potential energy of the electron in position 1 is eV_1 and in position 2 is eV_2. If v_1 and v_2 are the speeds of the electron at the points 1 and 2 respectively,

$$eV_2 - eV_1 = \tfrac{1}{2}m_e v_2^2 - \tfrac{1}{2}m_e v_1^2 \tag{1.6}$$

If the electron starts from rest and moves through a potential difference of V, it acquires a speed v given by

$$eV = \tfrac{1}{2}m_e v^2 \tag{1.7}$$

This equation is only valid when the speed v of the electron is small compared with the speed of light. In the above it is assumed that the mass m_e is constant. In fact, in accordance with the Special Theory of Relativity, the mass of a particle increases with its speed but this increase is only of major importance as the speed approaches that of light in free space. The correction can usually be neglected if the electron is accelerated through less than 5 kV.

Example 1.4

Calculate the speed of electrons which are accelerated from rest through a potential difference of $5\,kV$.

(Ratio of charge to mass, e/m_e, for the electron $= 1.76 \times 10^{11}$ coulomb per kilogramme).

From equation (1.7)

$$v = \sqrt{\frac{2Ve}{m_e}}$$

Substituting $V = 5000$ and $e/m_e = 1.76 \times 10^{11}\,\text{C}\,\text{kg}^{-1}$,
$v = \sqrt{(10\,000 \times 1.76 \times 10^{11})}$ metre per second
$= 4.2 \times 10^7 \text{m}\,\text{s}^{-1}$.

1.5 The Kinetic Theory of Gases

In Section 1.1. it is stated that electrons moving in an electric field within a metal are restricted in their motions by collisions with the metal atoms and that this motion has resemblance to that of molecules in a gas. It is necessary, therefore, to summarize relevant aspects of the kinetic theory of gases.

The following assumptions are made in the kinetic theory of ideal or perfect gases:

(a) The molecules of a particular gas are identical, hard, elastic spheres. The molecules of one gas differ from those of another only in mass.
(b) The molecules do not exert forces upon one another.
(c) The molecules travel along rectilinear paths between mutual collisions in a perfectly random fashion, i.e. all directions of motion are equally likely.
(d) The volume occupied by the molecules is negligible compared with that occupied by the gas.

For an ideal gas it is readily proved that

$$p = \tfrac{1}{3} mnc^2 = \tfrac{1}{3} \rho c^2 \tag{1.8}$$

where p is the pressure of the gas, m is the mass of a molecule, n is the number of molecules per unit volume (the molecular density), $\rho = mn$ is the density of the gas and c^2 is the mean square speed of the molecules.

The average kinetic energy of each molecule will be $\tfrac{1}{2}mc^2$ so the total kinetic energy E of the molecules contained in unit volume is $\tfrac{1}{2} mnc^2$. It is easily seen that equation (1.8) can therefore be written in the alternative form

$$p = \tfrac{2}{3} E \tag{1.9}$$

One kilogramme-molecule (i.e. one kilomole) of any substance whether in the solid, liquid or gaseous state consists of M kg of the substance where M is its molecular mass. The kilomole always contains N_A molecules where N_A is the Avogadro constant.

$$N_A = 6.03 \times 10^{26} \text{ per kilomole}$$

One kilomole of any gas at standard temperature and pressure [s.t.p., i.e. at 0°C and 1 atmosphere = 1.03×10^5 newton per square metre ($1.03 \times 10^5 \text{N m}^{-2}$)], occupies a volume of 22.4 cubic metre (22.4 m³).

If V is the volume occupied by one kilomole of gas, equation (1.9) can be written

$$pV = 2N_A E/3n \qquad (1.10)$$

because V is clearly equal to N_A/n. Experiment shows that

$$pV = RT \qquad (1.11)$$

where T is the absolute temperature and R is the gas constant, which has the same value for all gases. The gas constant R (sometimes called the kilomolar gas constant) can be defined from comparison of equations (1.10) and (1.11) as equal to two-thirds of the total energy of the molecules in one kilomole of gas at a temperature of 1K [$T = 1$ in equation (1.11)].

In equation (1.11) put $p =$ the standard atmospheric pressure = $1.013 \times 10^5 \text{ N m}^{-2}$, $V = 22.4 \text{ m}^3$ and $T = 273$ K (corresponding to 0°C). Then

$$R = \frac{1.013 \times 10^5 \times 22.4}{273}$$
$$= 8317 \text{ joule deg}^{-1}\text{K kilomole}^{-1}$$
$$= 8317 \text{ J K}^{-1}\text{kilomole}^{-1}$$

Boltzmann constant (k) equals R/N_A and may be defined as the gas constant per molecule.

$$k = R/N_A = 8317/(6.03 \times 10^{26}) = 1.38 \times 10^{-23} \text{J K}^{-1}$$

From equations (1.10) and (1.11)

$$2N_A E/3n = RT \qquad (1.12)$$

Hence $\qquad\qquad\qquad E/n = 3kT/2 \qquad (1.13)$

Hence $3kT/2$ is the kinetic energy of a single gas molecule at absolute temperature T.

Substituting the value for E in equation (1.9) gives

$$p = nkT \qquad (1.14)$$

Example 1.5(a)

Calculate the root-mean-square (rms) speed of hydrogen molecules at stp given that the density of hydrogen at stp is $9.0 \times 10^{-2} kg\ m^{-3}$. (Standard atmospheric pressure $= 1.013 \times 10^5\ N\ m^{-2}$).

From equation (1.8) $p = \tfrac{1}{3}\rho c^2$. Therefore

$$c = \sqrt{\left[\frac{3 \times 1.013 \times 10^5}{9 \times 10^{-2}}\right]} = 1.84 \times 10^3 m\ s^{-1}.$$

The idea of separation between molecules needs to be considered more specifically because in any gas at any pressure the molecules will not be regularly spaced. Indeed, *molecular chaos* is the natural order, implying that the molecules have all possible speeds (but a root mean square speed is quoted), all possible directions in space, and are separated by all possible distances, literally from zero to infinity. To be more specific, the idea of *mean free path* (λ) of a molecule is introduced: it is defined as the *average* distance that a molecule travels in a gas after hitting one molecule and before it collides with another molecule.

Regarding the molecule as a hard sphere of diameter d it can be shown that

$$\lambda = \frac{1}{\pi n d^2 \sqrt{2}} \tag{1.15}$$

where λ is the mean free path (m.f.p.) of a molecule of a gas moving among molecules of the same kind and n is the number of molecules per unit volume. This mean free path depends on the nature of the gas and for a given gas (i.e. given molecular diameter d) at a given temperature, it is seen from equations (1.15) and (1.14) that

$$\lambda = \frac{\text{const.}}{n} = \frac{\text{const.}}{p} \tag{1.16}$$

For nitrogen, a useful working rule is that

$$\lambda = \frac{6.65 \times 10^{-3}}{p}\ \text{metre} \tag{1.17}$$

where p is the pressure in $N\ m^{-2}$, so that at atmospheric pressure

$$\lambda = 6.65 \times 10^{-3}/(1.013 \times 10^5) = 6.6 \times 10^{-8}\ m \quad \text{approx.}$$

This is very nearly the same for air, so that the average separation between molecules in the atmosphere is very small unless the pressure is reduced considerably.

An electron moving in a gas will have a longer mean free path than a molecule because an electron is vanishingly small compared with a molecule (nominally the diameter of an electron is about 2×10^{-15}m compared with about 5×10^{-10}m for a molecule). The equation for the m.f.p. (λ_e) of an electron in a gas is

$$\lambda_e = \frac{4}{\pi n d^2} = 4\lambda\sqrt{2} \qquad (1.18)$$

The relationships derived in this section are not restricted to gas molecules but apply also to any assembly of particles which can be treated as minute elastic spheres. In particular they can be used in the discussion of the motion of free electrons among the atoms of an element. The electron theory of metals based on the kinetic theory of gases is called the classical model. It has been superseded by the statistical model developed by Fermi, Dirac and Sommerfeld and termed the FDS model.

Example 1.5(b)

The Boltzmann constant $k = 1.38 \times 10^{-23} J\ K^{-1}$. *On the basis that the electron-volt (eV) is defined as the work done in transferring an electron through a potential difference of one volt, express k in the unit $eV\ K^{-1}$ and calculate the mean energy in eV of an electron in a metal at 300 K. (The charge of the electron $= 1.6 \times 10^{-19}$ coulomb).*

As 1 volt = 1 joule per coulomb, therefore

$$1\ eV = 1\ J \times 1.6 \times 10^{-19} = 1.6 \times 10^{-19} J$$
$$k = 1.38 \times 10^{-23} J\ K^{-1}$$
$$= \frac{1.38 \times 10^{-23}}{1.6 \times 10^{-19}}\ eV\ K^{-1} = \frac{1}{11\ 600}\ eV\ K^{-1}$$

From equation (1.13) the mean kinetic energy of an electron is $E = 3kT/2$. Therefore at $T = 300$ K,

$$E = \frac{3 \times 300}{2 \times 11\ 600}\ eV = 0.038\ eV$$

Example 1.5(c)

The diameter of a nitrogen molecule is 3.2×10^{-10}m. Calculate the mean free path of nitrogen molecules at a pressure of (a) $10^5 N\ m^{-2}$ (approximately standard atmospheric pressure) and (b) at $2 \times 10^{-3} N\ m^{-2}$ (the gas pressure in a typical cathode ray tube). (The Avogadro constant $(N_A) = 6.03 \times 10^{26}$ kilomole^{-1}).

Equation (1.15) is $\lambda = 1/\pi n d^2 \sqrt{2}$. One kilomole of gas at s.t.p. occupies a volume of 22.4 m³. Therefore

$$n = \frac{6.03 \times 10^{26}}{22.4}\ \text{molecules/m}^3$$

(a) At a pressure of 10^5 N m^{-2},

$$\lambda = \frac{22.4}{\pi\sqrt{2} \times 6.03 \times 10^{26} \times (3.2)^2 \times 10^{-20}} \text{ m}$$
$$= 8.5 \times 10^{-5} \text{mm}$$

N.B. Although small, this mean free path is 265 times the molecular diameter, so the assumption given about the insignificance of the molecular volume compared with the gas volume is justified.

(b) For a given temperature, $\lambda \propto 1/p$ (equation (1.16)) so at a pressure of 2×10^{-3} N m^{-2} the mean free path must be $10^5/(2 \times 10^{-3})$ times that at 10^5 N m^{-2}. Hence

$$\lambda = \frac{10^5}{2 \times 10^{-3}} \times 8.5 \times 10^{-5}$$
$$= 4.25 \times 10^3 \text{mm} = 4.25 \text{ m}.$$

N.B. For an electron, from equation (1.18), moving in nitrogen at this pressure, the mean free path is $4.25 \times 4\sqrt{2} = 5.66 \times 4.25 = 24$ m. Such a low pressure therefore enables the electrons to traverse the cathode ray tube with a small probability that they collide with and ionize the gas molecules.

1.6 The Nuclear Model of the Atom: Ionization

The electrical and magnetic properties of materials are due primarily to the motions of electrons in the material. To understand the electrical properties of a solid it is consequently necessary to examine the atomic structure and also the arrangements of atoms within crystals.

The planetary or nuclear model of the atom, first described by Lord Rutherford in 1911, may be likened in some respects to our solar system. Around the positively charged nucleus, electrons rotate in certain orbits. These orbits are ellipses with the nucleus at one focus because the force of attraction between unlike charges (the positive nucleus and the negative electron) obeys the inverse square law.

In the neutral atom, the total number of electrons rotating in orbits around the nucleus is equal to the number of protons in the nucleus, which is Z, the atomic number. If an atom loses one or more of its orbital electrons, it is positively charged, and indeed becomes a positive ion. An atom which gains one or more electrons becomes a negative ion.

For ionization to occur, energy must be imparted to the atom to dislodge an electron from its outer orbital structure. This comes about by the incidence upon the atom of a sufficiently energetic particle. The chief ways of causing ionization are by the incidence of electrons and, less effectively, by irradiation with photons, the 'particles' of electromagnetic radiation which have an energy $h\nu$ where ν is the frequency of radiation and h is the Planck constant $(6.6 \times 10^{-34}$ joule second). Radiations from radioactive materials, in particular alpha-particles, beta-particles (high speed electrons) and, to a small extent, gamma rays, are also capable of causing ionization. Again, a material can be ionized by heating it to a sufficiently high temperature (e.g. gas in a flame).

The atoms of some solid semiconductors can be ionized by the incidence of visible light. These semiconductors are used to advantage in solid state photoelectric cells.

Example 1.6(a)
Calculate the energy in electron-volt of a photon in monochromatic light of wavelength 550 nm. (The speed of light in free space $= 3 \times 10^8 \text{m s}^{-1}$, *the Planck constant* $h = 6.6 \times 10^{-34} Js$).

The energy E of the photon of frequency ν is given by

$$E = h\nu = hc/\lambda$$

where c is the speed of light in free space and λ is the wavelength corresponding to a frequency ν. Substituting the values given

$$E = \frac{6.6 \times 10^{-34} \times 3 \times 10^8}{550 \times 10^{-9}} = 3.6 \times 10^{-19} \text{ joule}$$

As $1 \text{ eV} = 1.6 \times 10^{-19}$ joule (see Example 1.5(b)), therefore

$$E = \frac{3.6 \times 10^{-19}}{1.6 \times 10^{-19}} = 2.25 \text{ eV}$$

Example 1.6(b)
The radius of the electron orbit (assumed to be circular) in hydrogen atom (atomic number $Z = 1$) *is* 5.3×10^{-2} *nm. Calculate the speed with which the electron describes the orbit. (For the electron the mass is* $9.1 \times 10^{-31} kg$ *and the charge is* $1.6 \times 10^{-19} C$).

To retain the electron of mass m_e in a circular orbit of radius r with a peripheral speed v, a force of magnitude $m_e v^2/r$ is required to be directed towards the central nucleus. This is provided by the force of attraction between unlike charges. As the nuclear charge is Ze, where e is positive and numerically equal to the electron charge, therefore

$$\frac{Ze \cdot e}{4\pi\varepsilon_0 r^2} = \frac{m_e v^2}{r}$$

where ε_0 is the permittivity of free space. For the hydrogen atom $Z = 1$, and therefore

$$v = \sqrt{\frac{e^2}{4\pi\varepsilon_0 \, m_e r}}$$

The Structure of Matter and the Solid State

Substituting the values given together with $\varepsilon_0 = 8.854 \times 10^{-12}$ farad per metre,

$$v = \sqrt{\frac{(1.6 \times 10^{-19})^2}{4\pi \times 8.854 \times 10^{-12} \times 9.1 \times 10^{-31} \times 5.3 \times 10^{-11}}} \, m\,s^{-1}$$
$$= 2.2 \times 10^6 \, m\,s^{-1}.$$

Example 1.6(c)
Calculate the ionization potential of atomic hydrogen.

Ionization takes place when the electron is removed from the normal ground state (the state when it is as close to the nucleus as it can be) to outside the influence of the nucleus of the hydrogen atom; such detachment corresponds to removing the electron from a distance r (the radius of the circular orbit in the ground state) to infinity.

Assuming a point at an infinite distance from the nucleus to be at zero potential energy, an electron moving from infinity to a distance r from the nucleus of positive charge Ze requires an amount of work to be done, so acquires a potential energy E_p given by

$$E_p = \int_\infty^r \frac{Ze^2}{4\pi\varepsilon_0 r^2} \, dr$$

because work = force × distance and the force on an electron at a distance r is $Ze^2/4\pi\varepsilon_0 r^2$. Hence

$$E_p = -\left[\frac{Ze^2}{4\pi\varepsilon_0 r}\right]_\infty^r = \frac{-Ze^2}{4\pi\varepsilon_0 r}$$

Note that this potential energy E_p is negative. This is because the potential energy of an electron at an infinite distance from the nucleus is taken to be zero, and work has to be done *on* the electron to *remove* it from the attractive effect of the positive nucleus.

However, the electron of mass m_e also possesses kinetic energy E_K of $\frac{1}{2} m_e v^2$ where v is its peripheral speed which, at the distance r, is given from Example 1.6(b) to be

$$E_K = \tfrac{1}{2} m_e v^2 = Ze^2/8\pi\varepsilon_0 r$$

The total energy of the electron when in the circular orbit of radius r is therefore

$$E = E_K + E_p$$
$$= \frac{Ze^2}{8\pi\varepsilon_0 r} - \frac{Ze^2}{4\pi\varepsilon_0 r} = \frac{-Ze^2}{8\pi\varepsilon_0 r}$$

The negative sign arises simply from the fact that the reference point of zero potential energy is conventionally taken to be at infinity.

For hydrogen, $Z = 1$, therefore

$$E = e^2/8\pi\varepsilon_0 r$$
$$= (1.6 \times 10^{-19})^2/(8\pi \times 8.854 \times 10^{-12} \times 5.3 \times 10^{-11})$$

on substituting $e = 1.6 \times 10^{-19}$ C, $\varepsilon_0 = 8.854 \times 10^{-12}$ F m^{-1} and $r = 5.3 \times 10^{-11}$ m (see Example 1.6(b)). Hence

$$E = 2.2 \times 10^{-18} \text{J} = \frac{2.2 \times 10^{-18}}{1.6 \times 10^{-19}} \text{eV} = 13.7 \text{ eV}$$

because 1 eV = 1.6×10^{-19} joule. The ionization potential, defined as the minimum potential difference through which an electron must fall to acquire sufficient energy to ionize the material, is therefore 13.7 V for hydrogen.

1.7 Motion of Electrons in a Uniform Magnetic Field

Consider a beam of electrons of uniform speed v passing through a region of uniform magnetic flux density directed perpendicularly to the path of the beam (Figure 1.2). If there are n electrons each of charge e in a metre length of the beam each travelling with a speed v

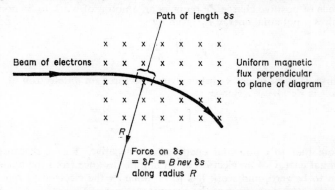

Figure 1.2. Passage of a beam of electrons of uniform speed through perpendicular uniform magnetic flux

in a vacuum, the rate of passage of charge at any point in the beam is nev, which is the current I. Therefore, the magnitude of the force δF due to the magnetic flux of density B on the electrons within a path length δs is given by

$$\delta F = BI . \delta s = Bnev . \delta s \quad (1.19)$$

So the force on each electron, of which there are $n\delta s$ in the length δs, is Bev. In accordance with Fleming's left-hand rule, this force is directed normally to the directions of both B and δs. If the beam makes an angle θ with the direction of the magnetic flux, the force on each electron becomes

$$F = Bev . \sin . \theta \quad (1.20)$$

The Structure of Matter and the Solid State

Three important considerations emerge from equation (1.20).

(a) When $v = 0$, $F = 0$; there is no force on a stationary electron in magnetic flux. In an electric field the electron would, on the other hand, be accelerated due to the force acting (Section 1.4).

(b) The force is a maximum when $\theta = 90°$, i.e. the electron moves at right angles to the direction of the magnetic flux, which is the direction of B. If $\theta = 0°$, $F = 0$ so there is no force acting on an electron moving along the magnetic flux lines.

(c) As the force is perpendicular to the velocity of the electron it can change the direction of motion but cannot affect its energy. Thus, an electron cannot gain energy from static magnetic flux.

For the special case of $\theta = 90°$ the force is always normal to the electron path which will cause the electron to move in a circular orbit at constant speed. Suppose this circle or arc of a circle in the magnetic field has a radius R (Figure 1.2). An electron travelling with constant speed v in a circular path of radius R has an acceleration towards the centre of the circle of v^2/R, the corresponding force being the mass times the acceleration, i.e. $m_e v^2/R$. This equals the force due to the motion of the charged particle in the magnetic flux, which is Bev.

$$Bev = m_e v^2/R \quad (1.21)$$

or
$$v = BeR/m_e$$

The time period T for one revolution of the electron is $2\pi R/v$. Hence

$$T = 2\pi m_e v/Bev = 2\pi m_e/Be \quad (1.22)$$

It is apparent that these equations (1.19) to (1.22) apply also to other charged particles than the electron where, in general, the electric charge is q and the mass is m.

1.8 The Arrangement of Electrons in Atoms: The Periodic Table

The elements occurring in nature range in atomic numbers from hydrogen ($Z = 1$) to uranium ($Z = 92$). Values for some of the intermediate elements of importance in this book are carbon ($Z = 6$), sodium ($Z = 11$), silicon ($Z = 14$), copper ($Z = 29$) and germanium ($Z = 32$).

An element such as copper is a very good conductor of electricity and has a positive temperature coefficient of resistance. This conductivity is due to the motion of electrons (within the copper) which move in an applied electric field. The atoms are within a face-centred cubic crystal structure. If the mechanism of this conduction is to be explained more intimately, the arrangement of the electrons in the electronic structure of the copper atom needs to be described more fully. Knowledge about this arrangement is also necessary to explain, for example, the semiconducting properties of silicon and why a compound such as aluminium oxide is an excellent insulator.

The experimental study of the line spectra of elements is the chief source of circumstantial evidence about atomic structure. All the evidence from spectroscopic and allied studies leads to the conclusion that each and every kind of atom has a specific structure.

For example, copper has an atomic number of 29 and a mass number (the nearest integer to its atomic mass) of 63 in its most abundant form. The nucleus of the copper atom thus has a positive charge of $29e$, where e is numerically equal to the electronic charge, due to the 29 protons in its nucleus. Within this nucleus there are also $(63 - 29) = 34$ neutrons. Around the nucleus in the neutral atom circulate 29 electrons.

The proton and the neutron have nearly equal masses and each are nearly 2000 times the mass of the electron. The massive very tiny central nucleus will therefore not be subject to any significant disturbance in the process of electric conduction or the emission of spectral radiation, ordinary chemical reactions and so on. It is to the electrons outside the nucleus that we must look to explain such phenomena.

In the copper atom, for example, each and every one of the 29 electrons circulating the nucleus has a most probable location with respect to the nucleus. Is there any manner in which these locations can be specified? The answer is a qualified yes. A vast amount of experimental evidence about the behaviour of the atoms of the elements can be explained on the basis that each electron in the atom is in a specific state (fully designated in the Pauli exclusion principle by a set of four quantum numbers, which will not be discussed here) and that no two electrons in a given atom can be in the same state. These concepts lead to the valuable picture of the atom in which electrons are in specific groups called main *shells* and *sub-shells* where the sub-shell is a part of a main shell. The main shells, numbered $n = 1, 2, 3, 4, 5$ proceeding 'outwards' from the nucleus are called successively the K, L, M, N, and O shells respectively. By 'outwards' is meant in a simple model of the atom, further from the nucleus More specifically, the shells of lower number n (called the principal quantum number) are the more closely bound to the nucleus and electrons in them have to receive more energy to dislodge them from the atomic structure. Thus, in spectroscopy, photons of energies in the x-ray region have to be used to dislodge electrons from the K and L shells of copper whereas those in the outermost shells are 'free' to take part in electrical conduction and in the formation of copper compounds such as the oxide.

The Structure of Matter and the Solid State

For a given main shell number n, it is found that the maximum number of electrons it can contain is $2n^2$. Furthermore, when a given shell contains its full complement of electrons it is said to be 'closed' or 'filled'. A filled shell forms a particularly stable arrangement of electrons, not easily disturbed by external influences. For example, for $n = 2$, $2n^2 = 8$. If this second shell *does* contain 8 electrons it is extra-stable and electrons are not so readily extracted from it to take part in physical and chemical phenomena as if it were to contain less than 8 electrons. Thus we get Table 1.1.

Main shell number n	1	2	3	4	5
Letter designation	K	L	M	N	O
Number of electrons in filled shell	2	8	18	32	50

Table 1.1

The familiar element sodium ($Z = 11$) will have 11 electrons outside the nucleus in its atom. The innermost shells will be filled because the electrons will occupy allowed states as closely bound to the nucleus as possible. For sodium, the K shell and the L shell will be filled with 2 and 8 electrons respectively, leaving the eleventh electron in an unfilled M shell. This lonely outermost electron is at the mercy of external forces applied to sodium. It is the electron responsible for the conduction of sodium, for the visible emission spectrum (the familiar lines at 589 and 589.6 n m in the sodium spectrum) and is the electron which is readily transferred to another element in the formation of a chemical compound (e.g. sodium chloride) so it is the *valence electron*.

The sub-shells are designated by a second quantum number l which, like n, is also an integer.* Within a main shell n, the possible values of l are 0, 1, 2, 3 etc. up to $(n - 1)$. Again, designation by letter notation is used. The lower case letters employed are s, p, d, f, g corresponding to l values of 0, 1, 2, 3, 4 respectively. These letters are chosen for somewhat unsatisfactory historical reasons arising from the early days of observations on the emission line spectra of elements: they are respectively the initial letters of the descriptive terms *sharp, principal, diffuse* and *fundamental*; beyond f, further letters are in alphabetical order.

A sub-shell characterized by an integer l is filled when it contains $2(2l + 1)$ electrons. For example, the letter d corresponds to $l = 2$;

---

* This assumption has to be modified in more advanced theory.

a d sub-shell is filled when it contains $2(2 \times 2 + 1) = 10$ electrons. Filled sub-shells are also stable configurations within the main shells though their stabilities are not so marked as those of main shells.

On this basis Table 1.1. can be extended to give Table 1.2. This table which gives the maximum numbers of electrons that can exist in the shells and sub-shells of atoms (based on the Pauli exclusion principle which is at the heart of the considerations which lead to the numbers quoted), is of the utmost value because it forms the basis of the explanation of the electronic structure of the elements in the periodic table.

Main shell (letter)	K	L		M			N				O				
Main shell number (n)	1	2		3			4				5				
Number of electrons in filled shell ($2n^2$)	2	8		18			32				50				
Sub shells within main shell (l)	0	0	1	0	1	2	0	1	2	3	0	1	2	3	4
Letter designation of l	s	s	p	s	p	d	s	p	d	f	s	p	d	f	g
Number of electrons in filled sub-shell $[2(2l+1)]$	2	2	6	2	6	10	2	6	10	14	2	6	10	14	18

Table 1.2

In Table 1.3 is given such detail for the first 37 elements from hydrogen ($Z = 1$) to rubidium ($Z = 37$). The horizontal lines on this table mark the closure of a shell or sub-shell arrangement.

In relation to this Table 1.3 note the closed shell or sub-shell structure corresponding to the inert gases helium ($Z = 2$), neon ($Z = 10$), argon ($Z = 18$) and krypton ($Z = 36$).

Some of the elements to which reference will be made later in this text are sodium, copper, aluminium, carbon, silicon, germanium, chlorine and oxygen. Consider these elements in relation to Table 1.3:

Sodium ($Z = 11$): a single electron in the 3s state (main shell $n = 3$; sub-shell $l = 0$, corresponding to s) outside a filled shell structure like that of neon. This single electron is the valence electron; it is also the electron which is 'free' to take part in electrical conduction. When it leaves the atom, the remaining structure is a positive ion; sodium is an electropositive element of valence unity, i.e. is monovalent.

Copper ($Z = 29$): a single electron in the 4s state outside filled shells K (2 electrons), L (8 electrons) and M (18 electrons). This

The Structure of Matter and the Solid State

Element	Atomic no. Z	K 1s	L 2s	L 2p	M 3s	M 3p	M 3d	N 4s	N 4p	N 4d	N 4f	O 5s	O 5p	O 5d	O 5g
H	1	1													
He	2	2													
Li	3	2	1												
Be	4	2	2												
B	5	2	2	1											
C	6	2	2	2											
N	7	2	2	3											
O	8	2	2	4											
F	9	2	2	5											
Ne	10	2	2	6											
Na	11				1										
Mg	12		Core		2										
Al	13		of 10		2	1									
Si	14		electrons		2	2									
P	15		as in		2	3									
S	16		neon		2	4									
Cl	17				2	5									
A	18				2	6									
K	19							1							
Ca	20							2							
Sc	21						1	2							
Ti	22		Core				2	2							
V	23		of 18				3	2							
Cr	24		electrons				5	1							
Mn	25		as in				5	2							
Fe	26		argon				6	2							
Co	27						7	2							
Ni	28						8	2							
Cu	29						10	1							
Zn	30						10	2							
Ga	31						10	2	1						
Ge	32						10	2	2						
As	33						10	2	3						
Se	34						10	2	4						
Br	35						10	2	5						
Kr	36						10	2	6						
Rb	37	Core of 36 electrons as in Krypton										1			

Table 1.3

Arrangements of electrons in the main shells and sub-shells of atoms 'free' electron, shielded from the nucleus by 28 electrons, is the main conduction electron. (Figure 1.3).

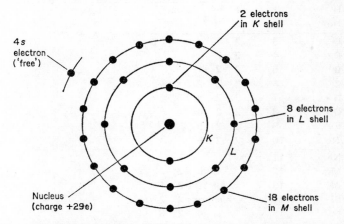

Figure 1.3. Electrons in shells in the copper atom (Schematic diagram)

Aluminium ($Z = 13$): three electrons outside a filled shell structure like that in neon. Two of these electrons are in the $3s$ state and one in $3p$. A good conductor which exhibits trivalent activity when all these three electrons enter in chemical combination, as in aluminium oxide, Al_2O_3.

Carbon ($Z = 6$): four electrons in the L shell outside the filled K shell. These four are half the full complement possible. Carbon is intermediate in electrical and chemical behaviour between electropositive metallic lithium and the electronegative gas fluorine ($Z = 9$) which requires only one electron to complete the outermost shell. Carbon has semiconducting properties and, like silicon and germanium, a negative temperature coefficient of resistance. Its maximum valence is four, i.e. quadravalent.

Silicon ($Z = 14$) and *germanium* ($Z = 32$): similar to carbon because the first has four electrons outside closed K and L shells and the second has four electrons outside filled K, L and M shells. These elements are of prime importance in semiconducting devices, in particular the semiconductor diode and the transistor.

Chlorine ($Z = 17$) and *fluorine* ($Z = 9$): both need only one electron to fill their shell structures. They are elements which avidly acquire electrons to become negative ions both in gases and in crystalline solids. Both are monovalent electronegative elements. Both combine

The Structure of Matter and the Solid State 21

very readily with monovalent electropositive alkali metals such as lithium, sodium and potassium.

Oxygen ($Z = 8$): requires two more electrons to complete its unfilled L shell. It is usually a divalent electronegative element, tending to form negative ions in gases and solids.

1.9 Simple Application of Electron Gas Theory

Consider the conduction of the metal sodium. As shown in Section 1.8 the sodium atom has a single electron in the M shell outside 10 electrons in closed K and L shells. These ten electrons provide an electrostatic screen between this outermost single electron and the positive nucleus with its charge of $11e$. This single electron in the unfilled M shell in an atom amongst myriads of atoms in a piece of sodium has no more definite affinity to any one atom than it does to the immediately neighbouring atoms. It is said to be 'free'. When an electric field is applied to sodium any free electrons therefore move readily along the electric field lines. In this respect they move in the space between the atoms in the crystal lattice of sodium metal (actually a body-centred cubical lattice) rather like molecules move in the space between molecules in a gas. The analogy can be carried further in that the restriction to the motion of the free electrons in sodium (and other metallic conductors) is brought about by collisions of these electrons with sodium atoms in the crystal lattice, comparing with the restriction on molecular motion in gases due to collisions between molecules.

In an early attempt to examine the mechanism of conduction in a metal, the following assumptions were made:

(*a*) The free electrons collide repeatedly with atoms in the metal.
(*b*) These collisions are elastic.
(*c*) The random velocity v of free electrons at an absolute temperature T is given by

$$\tfrac{1}{2}m_e v^2 = 3/2kT \tag{1.23}$$

where k is the Boltzmann constant and m_e is the mass of the electron.

Equation (1.23) compares with equation (1.13) in the kinetic theory of gases.

The mean drift velocity u that the free electrons acquire in the direction of the electric field is very small compared with their random velocities due to thermal agitation of which the mean square value is given by equation (1.23).

Applying the concept of mean free path λ to free electrons in a metal (*cf.* Section 1.5 on the kinetic theory of gases), an electron is free to drift in the electric field for a time t between collisions with lattice atoms where

$$t = \lambda/v \tag{1.24}$$

Note that we have v in this equation (1.24) and *not* u; the time of flight of electrons between collisions is decided by the thermal agitation velocity and not the much smaller drift velocity.

The force Ee on each free electron of mass m_e in a field E will produce an acceleration of Ee/m_e. Immediately after a collision between a free electron and the much heavier atom in the lattice, the electron velocity will be negligibly small and is taken to be zero. During the subsequent time t before it collides with another atom it will acquire a velocity given by acceleration multiplied by time, i.e. Eet/m_e. The average drift velocity can be taken to be the mean of the velocity just after a collision and just before a collision, so that

$$u = Eet/2m \tag{1.25}$$

If there are n free electrons per unit volume in the metal, the current density J (the current per unit area of cross-section drawn perpendicular to the applied electric field E) is given by

$$J = neu \tag{1.26}$$

Substituting for u from equation (1.25) gives

$$J = ne^2Et/2m \tag{1.27}$$

Substituting for t from equation (1.24), therefore,

$$J = ne^2E\lambda/2mv \tag{1.28}$$

But $v = \sqrt{(3kT/m)}$ from equation (1.23). Hence

$$J = ne^2E\lambda/2\sqrt{(3kTm)} \tag{1.29}$$

Now the conductivity $\sigma = 1/\rho$, where ρ is the resistivity. But $\rho = RA/l$ where R is the resistance of a specimen of cross-sectional area A and length l. Therefore

$$\sigma = l/RA$$

If a p.d. of V exists across a specimen of length l, the electric field strength $E = V/l$. If the resulting current is I then, from Ohm's law,

$$I = V/R = VA\sigma/l$$

Hence
$$\sigma = \frac{Il}{VA} = \frac{I/A}{V/l} = \frac{J}{E}$$

Equation (1.29) therefore gives

$$\sigma = ne^2\lambda/[2\sqrt{(3kTm)}] \quad (1.30)$$

Example 1.9

The resistivity of sodium at $0°C$ (273 K) is $4.3 \times 10^{-8} \, \Omega \, m$. Given that there are 2.6×10^{28} free electrons per cubic metre of sodium (calculated on the basis that there is one free electron per atom) calculate the mean free path of electrons in sodium. (Electron charge $e = 1.6 \times 10^{-19} C$; electron mass $m_e = 9.1 \times 10^{-31} kg$; Boltzmann's constant $k = 1.38 \times 10^{-23} \, J\,K^{-1}$).

From equation (1.30) the electrical conductivity σ is given by

$$\sigma = ne^2\lambda/[2\sqrt{(3kTm_e)}]$$

Substituting $\sigma = 1/(4.3 \times 10^{-8})$ and the other values given,

$$\lambda = \frac{2[3 \times 1.38 \times 10^{-23} \times 273 \times 9.1 \times 10^{-31}]^{\frac{1}{2}}}{4.3 \times 10^{-8} \times 2.6 \times 10^{28} \times (1.6 \times 10^{-19})^2}$$
$$= 7 \times 10^{-9} m = 7 \, nm.$$

1.10 Limitations of the Electron Gas Model of Electrical Conduction in Metals

The maximum lattice spacing of sodium is determined by x-ray crystallography to be 0.43 nm. The value for the mean free path of the electron in sodium as calculated in Example 1.9 is nearly twenty times this spacing. This disparity indicates the weakness of this simple theory of the electron gas model of metallic conduction.

The analysis given in Section 1.9 is successful in that the result expressed by equation (1.30) is in accordance with experimental determinations of σ (which depend on Ohm's law) in that the conductivity is a constant at a constant temperature T and independent of the applied field E. However, equation (1.30) indicates that σ is proportional to $T^{-\frac{1}{2}}$ for a given metal at a given temperature whereas experiment shows that, to a close approximation, σ is inversely proportional to T, i.e. proportional to T^{-1}.

The limitations of the simple electron gas model are thus apparent and further experimental results in other contexts give rise to yet greater doubt about its veracity. A further success of this simple model was, however, that it could be used to calculate the thermal conductivity K of a metal. Experience has shown that a good electrical conductor is also a good thermal conductor. Furthermore, the ratio of the coefficient of thermal conductivity K to the coefficient of electrical conductivity σ is approximately a constant for all metals at room temperature. This relationship is expressed in the Wiedemann–Franz law of 1853,

$$K/\sigma = \text{const.} \times T \quad (1.31)$$

where T is the absolute temperature. By assuming that free electrons are responsible for thermal energy an equation similar to (1.30) may be derived.

It is now known that the energies of the free electrons in a metal are not distributed in the same way as the energies of molecules in a gas. Whereas the molecules in a gas have an energy distribution given by a function due to Maxwell and Boltzmann, the distribution function for electrons in a metal is that given by Fermi and Dirac. Nevertheless, the free electron gas model modified by the use of Fermi–Dirac statistics in place of Maxwell–Boltzmann statistics is still of considerable value.

1.11 Some Electrical Properties Characteristic of Metals

(a) Metals have resistivities within the range 10^{-8} to $10^{-5}\Omega$ m. Silver, one of the best conductors, has a resistivity of $1.5 \times 10^{-8}\Omega$ m at room temperature whereas that of manganese is $2.6 \times 10^{-6}\Omega$ m.

(b) Metals have a very high electron density which does not change with temperature; thus the number of current carriers (free electrons) available is constant.

(c) Metals have a small positive temperature coefficient of resistivity of approximately $0.3\% \ K^{-1}$. A rise in temperature increases the amplitude of vibration of the lattice atoms which consequently retard more the drift motion of the free electrons in an applied electric field. The resistance of a metal therefore increases with temperature.

(d) The resistivities at room temperature of any two polycrystalline samples of the same metal are virtually identical even though the polycrystalline structures differ.

1.12 Semiconductors

For about 100 years, solids have been known with resistivities in the range 10^{-3} to $10^4\Omega$ m. It is within this group that semiconductors exist. However, while some impure or heavily doped semiconductors can have electrical properties quite similar to metals, those in a very pure state may be good insulators, especially if maintained at low temperature.

Much of the early work on semiconductors carried out over many years prior to the discovery of the transistor produced very confusing results. The measured electrical properties of two apparently identical specimens were often quite different. At the time the influence of impurity atoms, the surface state of the sample and structural imperfections were not appreciated. Thus structural

The Structure of Matter and the Solid State

imperfections, in particular the grain boundaries between the tiny crystals of a polycrystalline specimen, interfere with the electron motion and so radically alter the electrical behaviour. It was not until the specimens were cut from large nearly perfect crystals that consistent and reproducible experimental results were obtained.

Some semiconductors of importance in electronics are listed below. Although some metallic salts, particularly the alkali metal halides (e.g. NaCl) show moderate ionic conductivity at high temperatures—which classifies them as semiconductors—they are at present of little interest in electronics so will not be discussed.

Elements Boron (B), silicon (Si), germanium (Ge), tellurium (Te), selenium (Se).

Compound semiconductors These are also called 'inter-metallic compounds' or '3-5' compounds. They are compounds of metals of which the component metals come from the third and fifth groups of the periodic table. Two much used examples are gallium arsenide (GaAs) and indium antimonide (InSb). Thus gallium has an atomic number Z of 31. It therefore has three electrons outside 28 electrons in the filled K, L and M shells which contain 2, 8 and 18 electrons respectively (Table 1.3). As a trivalent element it is group III of the periodic table. Arsenic has an atomic number Z of 33; it therefore contains 5 electrons outside the 28 electrons in the filled K, L and M shells (Table 1.3) and so is in the group V of the periodic table.

For reference purposes it should be noted that group III is divided into group IIIA containing the elements boron, aluminium, *gallium*, *indium* and titanium and group IIIB comprising scandium, yttrium, lanthanum and actinium; also group V consists of group VA (nitrogen, phosphorus, *arsenic*, antimony and bismuth) and group VB (vanadium, niobium and tantalum).

Note also indium in group III and antimony in group V. It is also significant that the three valence electrons of a metal in group III combine with the five electrons in a metal of group V to provide the eight electrons of a relatively stable shell structure.

Other inorganic compounds In particular we have cadmium sulphide (CdS), zinc sulphide (ZnS) and zinc oxide. Cadmium ($Z = 48$) and zinc ($Z = 30$) are both divalent metals in group II of the periodic table (e.g. zinc contains two electrons in the N shell outside the twenty-eight in the filled K, L and M shells, see Table 1.3) whereas sulphur ($Z = 16$) and oxygen ($Z = 8$) are in group VI (e.g. sulphur contains six electrons outside filled K and L shells, see Table 1.3). These semiconducting inorganic compounds are hence known as '2-6' compounds.

Apart from a conductivity within a certain range, semiconductors usually show electrical behaviour summarized as follows:

(i) A large negative temperature coefficient of resistance of approximately 6% K^{-1}.

(ii) Sensitivity to light: incident radiation produces a photovoltaic or photoconductive effect.

(iii) The electrical properties change markedly when the specimen is placed in a magnetic field. The current carriers are strongly influenced so that magnetoresistance (resistance change when placed in a magnetic field) and the Hall effect (Section 1.20) are much larger than for metals.

(iv) Impurity atoms have a very marked effect. For example, a pentavalent impurity (group V element) added to the extent of only 1 part in 10^8 will increase the conductivity of germanium by 10 times at 300 K.

(v) Properties are considerably affected by the surface state of the sample. Molecules of water on the surface of a crystal specimen alter profoundly its electrical behaviour. Also when a contact is made between a metal and a semiconductor, it may be ohmic (i.e. obey Ohm's law) or have rectifying properties depending on the state of the interface at the contact. An ohmic contact will have linear characteristics, so that the current through the junction will be a linear function of the voltage across it. Ohmic contacts are often made by doping heavily the semiconductor where the metal is to be attached.

The electron gas model applied with modest success to metals (Section 1.9) is of little value in explaining the electrical behaviour of semiconductors. One of the triumphs of the quantum theory of solids in the 1930s was the explanation of the effects of current carriers in semiconductors. This work led to the energy band theory of solids which not only explained the temperature and impurity sensitivity of semiconductors but also enabled materials to be broadly classified into conductors, semiconductors and insulators in terms of their electronic or extra-nuclear atomic structure.

1.13 Intrinsic Semiconducting Elements

The two important elements are silicon ($Z = 14$) and germanium ($Z = 32$). These are alike in that silicon has 4 valence electrons outside filled K and L shells (2 in K and 8 in L) whereas germanium has 4 valence electrons outside filled K, L and M shells (18 in M). They are both quadravalent elements and are in group IV (strictly

IVA) of the periodic table in which also occurs carbon ($Z = 6$) which has four valence electrons outside the two electrons in the filled K shell.

A crystalline form of carbon is diamond. Silicon and germanium both crystallize with a diamond-like structure. In the crystal lattice of germanium (represented in two dimensions in Figure 1.4) each of the four valence electrons of a particular atom is shared by one of the four nearest neighbouring atoms. With all four of the valence electrons used in the covalent bonding (as it is called) there are no

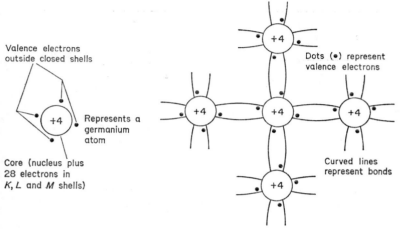

Figure 1.4. Atoms of germanium in the crystal lattice showing the covalent bonds between neighbouring atoms. (A two-dimensional schematic diagram of what is actually a three-dimensional array)

free electrons available. At low temperature (near 0 K) germanium (and silicon) is therefore an insulator.

If the temperature of a germanium crystal is raised, the electrons gain energy from the lattice vibrations which increase with temperature. Some of the valence electrons may gain sufficient energy to break free from the covalent bonds and become available as current carriers.

An electron so released from such bondage must leave behind a vacancy in the lattice. Such a vacancy is called a *hole*. This hole therefore exists within the valence electron structure of the atom. It cannot exist for any significant time because the positive nucleus of the atom concerned will attract an electron (able to move in an applied electric field) from elsewhere in the neighbouring atoms so the hole becomes filled, the normal valence structure of the atom becoming completed thereby. This must transfer the vacancy (hole)

28 An Introduction to Semiconductors

to the neighbouring atom which, in turn, passes its electron deficiency to the next neighbour, and so on. The hole is therefore mobile in an applied electric field, but hole motion will take place in *the opposite direction to electron motion*. The hole therefore has the property of a positive charge, and is known as a *positive hole*. Electron and positive hole motion in an applied electric field are illustrated schematically by Figure 1.5. It should be noted that the concept of positive hole conduction is a convenience; the reality is still the motion of electrons.

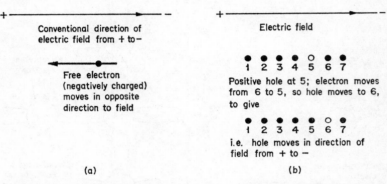

Figure 1.5. (a) Electron motion in an applied electric field and (b) positive hole motion in a semiconductor in an applied electric field

The semiconducting crystalline elements silicon and germanium in the pure state therefore exhibit electrical conductivity at elevated temperatures because of the generation within their lattices of *electron-hole pairs* where the electron is freed from covalent bonding by temperature rise and necessarily leaves behind it the hole or vacancy. The conductivity resulting from the generation of electron-hole pairs in pure silicon or germanium, free from lattice imperfections, is known as *intrinsic conductivity*. It is distinguished from *extrinsic conductivity* which results (Section 1.14) from impurities in the crystal.

Intrinsic conductivity may be attributed to ionization produced thermally in the solid silicon or germanium. At any instant electrons and holes are being produced and are also recombining with one another. At any given temperature, a state of dynamic equilibrium exists between the ionization and recombination processes.

1.14 Extrinsic or Impurity Conductivity

The conductivity of germanium (and of silicon) can be increased, while still preserving the crystal lattice configuration, by the addition of certain impurities. Atoms are chosen which can fit in to the lattice structure without inducing undue strain. The most useful impurities to introduce in a controlled fashion are selected elements with a valence of 5 (pentavalent) or 3 (trivalent), that is either one more or one less than the quadravalent germanium. The addition of such impurities is known as *doping*.

Pentavalent (i.e. group V) elements suitable for doping the germanium or silicon crystal are arsenic (As), phosphorus (P) or antimony (Sb). Reference to Table 1.3 shows, for example, that arsenic ($Z = 33$) has 5 valence electrons outside the 28 electrons in the filled K, L and M shells which, together, contain $2 + 8 + 18 = 28$ electrons.

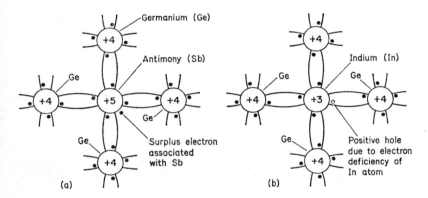

Figure 1.6. (a) One atom of germanium in the crystal lattice is replaced by pentavalent antimony leaving a surplus free electron and (b) one atom of germanium is replaced by trivalent indium leaving a surplus positive hole

The effect of incorporating pentavalent antimony within the germanium lattice is shown in Figure 1.6(a). Four of the five valence electrons of antimony will take part in the covalent bonds. The fifth valence electron is bound only very weakly to its parent atom. Only a small amount of energy is required to free this fifth electron, indeed only about 0.01eV as compared with the 0.75eV required in pure germanium.

Before the temperature is high enough for appreciable intrinsic conductivity of the germanium itself to occur, all the impurity atoms will have contributed a free electron and become ionized. The pentavalent antimony donates conduction electrons to the germanium so is called a *donor impurity*. The impure or doped germanium is called *n-type* because negative charges (i.e. electrons) are responsible for the increased conductivity consequent on pentavalent impurity additions.

On the other hand, if the added impurity is trivalent, for example, aluminium, boron or indium, the modified lattice structure is as shown in Figure 1.6(b). Aluminium ($Z = 13$), for example, is trivalent in group III of the periodic table because it has three valence electrons outside a closed shell structure of 10 electrons comprising 2 in the filled K shell and 8 in the filled L shell.

As an example, the widely used trivalent element indium ($Z = 49$) has one too few valence electrons to provide the four necessary for the covalent bonds in quadravalent germanium or silicon. There is consequently a vacancy or positive hole left in the lattice. This hole readily captures a thermally energized electron from a neighbouring atom and again this transfer requires less energy than that required for an electron to break free from the germanium bond.

Enhanced conduction is now primarily due to positive holes and the doped germanium (or silicon) is called *p-type*. The added impurity is known as an *acceptor* because it accepts an electron to to fill the vacancy and complete the bonding.

A donor atom becomes a positive ion when it donates an electron whereas an acceptor atom becomes a negative ion when it gains an electron. Nevertheless, the total charge on the crystal is always zero because the ionization merely results from the production of electron-hole pairs. Ions formed in a doped crystal are locked in the lattice, otherwise they too would contribute to the conductivity. Ionization can be produced by the action of light and this process is used to advantage in the photodiode and the phototransistor (Section 5.15). Transistors are mounted inside light-tight capsules to exclude this interference.

Purity in semiconductors is a relative term. A sample of germanium which has been subjected to zone refining techniques (Chapter 2) might have an impurity concentration of 1 part in 10^{10}, corresponding to 10^{19} impurity atoms per cubic metre. This is termed a 'near-intrinsic specimen', although it will exhibit very low n- or p-type conductivity depending upon whether the residual impurity is pentavalent or trivalent. A doped n-type material will have electrons as *majority carriers*. Because some small quantity of trivalent impurity is inevitably present, some hole conduction must occur. These positive holes are called *minority carriers*.

It is possible to convert n-type to p-type or *vice versa* simply by adding sufficient impurity to outnumber the majority carriers, which will then be in the minority. In manufacturing processes, the material is refined to a near-intrinsic state and the impurity added to achieve a conductivity of the correct type and magnitude.

1.15 Carrier Mobility and Current Density

The motion of current carriers (electrons and positive holes) can occur as the result of

(a) drift in an applied field;
(b) diffusion from a region of high concentration to one of lower concentration.

Only the first of these processes is examined at present. At any given temperature the carriers move in random directions with a distribution of speeds and make collisions with the vibrating lattice ions. When an electric field is applied to the semiconductor, positive holes acquire a drift component velocity in the direction of the field whereas electrons drift in the opposite direction (this is using the convention that the direction of an electric field is that in which a positive charge would move or tend to move). Superimposed on the random motion, therefore, is a velocity component produced by the applied field.

If n_p is the density of positive holes (number of positive holes per cubic metre) and v_p metre per second is their drift velocity, the current density J_p, defined as the current through unit cross-sectional area perpendicular to v_p, due to positive holes is given by

$$J_p = n_p e v_p \text{ ampere per square metre} \qquad (1.32)$$

where e is the electronic charge in coulomb. Similarly, the current density J_n resulting from electron motion in the direction of the applied field is given by

$$J_n = n_n e v_n \text{ A m}^{-2} \qquad (1.33)$$

where the subscript n denotes electron. The mobility k of a current carrier is defined as the velocity acquired in unit electric field. Therefore

$$k_n = v_n/E \quad \text{and} \quad k_p = v_p/E \qquad (1.34)$$

the unit of mobility k being (metre per second)/(volt per metre), i.e. m s^{-1}/V m^{-1} or m^2 V^{-1} s^{-1}.

The total current density J is simply the sum of J_n and J_p. Hence, from equations (1.32), (1.33) and (1.34),

$$J = eE(n_n k_n + n_p k_p) \qquad (1.35)$$

The conductivity σ is the current density in unit electric field, so

$$\sigma = e(n_n k_n + n_p k_p) \qquad (1.36)$$

32   *An Introduction to Semiconductors*

For a near intrinsic specimen, equal numbers of electrons and holes exist at any instant of time so $n_n = n_p = n_i$, where n_i is the density of current carriers in the intrinsic case. Equation (1.36) can then be written

$$\sigma = n_i e(k_n + k_p) \tag{1.37}$$

Example 1.15(a)
In a near intrinsic specimen of germanium at 300 K the electron density is 2×10^{19} m^{-3}. If the mobilities of the electrons and positive holes are respectively 0.39 and 0.19 m^2 $V^{-1} s^{-1}$, calculate the resistivity of germanium.

Using equation (1.37),

$$\sigma = 2 \times 10^{19} \times 1.6 \times 10^{-19}(0.39 + 0.19)$$

on putting the electronic charge $e = 1.6 \times 10^{-19}$ coulomb. Hence

$$\sigma = 1.86 \text{ S m}^{-1}$$

(S = siemen, the reciprocal of ohm.)

$$\text{Resistivity } \rho = 1/\sigma = 1/1.86 = 0.54 \text{ } \Omega \text{ m}.$$

Example 1.15(b)
Calculate the mobility and mean free time at $0°C$ of electrons in silver given that there are 10^{29} free electrons per cubic metre in silver and its resistivity at 273 K ($0°C$) is 1.5×10^{-8} Ω m.

The current density J in a silver specimen is given by

$$J = nev$$

where n is the current carrier density, and only electrons act as current carriers, e is the electronic charge and v is the drift velocity of the electrons in an electric field E. Therefore

$$\sigma = \frac{J}{E} = \frac{1}{\rho} = nek_n$$

where the conductivity is σ, the resistivity is ρ and k_n is the mobility of the electrons. Hence

$$k_n = 1/ne\rho = \frac{1}{10^{29} \times 1.6 \times 10^{-19} \times 1.5 \times 10^{-8}}$$

on substituting the values given and $e = 1.6 \times 10^{-19}$C. Thus

$$k_n = 4.2 \times 10^{-3} \text{ m}^2 \text{ V}^{-1} \text{ s}^{-1}$$

The mean free time occupied by the electrons between collisions with silver atoms in the lattice is given by equation (1.25):

$$u_n = Eet/2m_e$$

where u_n is the mean drift velocity of the electrons and m_e is the mass of an electron. Hence

$$t = \frac{2m_e u_n}{Ee} = \frac{2m_e k_n}{e}$$

because $u_n = k_n E$. Substituting for k_n as calculated above and putting $m_e = 9.1 \times 10^{-31}$ kg,

$$t = \frac{2 \times 9.1 \times 10^{-31} \times 4.2 \times 10^{-3}}{1.6 \times 10^{-19}}$$
$$= 4.8 \times 10^{-14} \text{s}.$$

1.16 Energy Levels in Isolated Atoms and Energy Bands in Solids; Metallic Conduction

In the gaseous state the mean separations (the mean free paths) between neighbouring atoms are several hundred times the diameters of the atoms themselves (*cf.* Example 1.5(c)) unless the gas pressure is many times greater than atmospheric pressure. The electrons around the nucleus, and even those farthest from the nucleus, in an atom in a gas are therefore relatively remote from those of the neighbouring atoms. Interactions between the outermost electronic structures of neighbouring atoms are negligible. From this point of view, the atoms (or molecules) in a gas may be said to be isolated from one another.

In the solid state, on the other hand, the separations between neighbouring atoms are comparable with the atomic diameters. Now the electronic structure of any one atom *will* interact with that of its neighbour atoms—the atoms are *not* isolated from one another.

Each and every electron in the extra-nuclear (outside the nucleus) structure of an isolated atom will occupy a discrete energy level, and these levels are identical with those of any other of the atoms. For the simplest atom—that of hydrogen—the main structure of the energy level diagram is shown in Figure 1.7. This data is obtainable by experiment (the study of the emission line spectra being the most useful technique) and is predicted with considerable accuracy by the theory of the hydrogen atom due to Niels Bohr (1913).

If atoms are close together in a solid, the atoms interact and the sharp electron energy levels (as represented for the hydrogen atom in Figure 1.7 with similar but more complex diagrams for the other isolated atoms) broaden into energy bands.

The simplest element which can exist at ordinary temperature in both the vapour and solid states is lithium (melting point 186°C). This has an atomic number Z of 3; there are two electrons in the filled K shell and one electron (the valence electron) in the L shell. For lithium vapour the sharp energy levels are shown in Figure 1.8(a). There are two electrons in the K shell with very nearly the same

34   *An Introduction to Semiconductors*

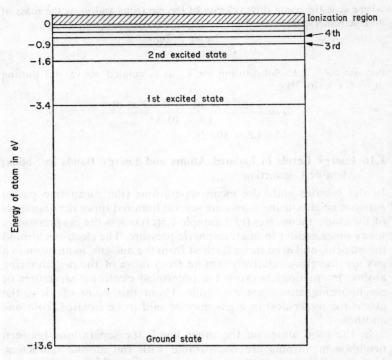

Figure 1.7. Energy-level diagram for the hydrogen atom

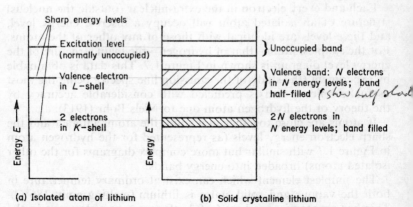

(a) Isolated atom of lithium

(b) Solid crystalline lithium

Figure 1.8. (a) Energy levels in isolated lithium atoms and (b) energy bands in a lithium crystal

The Structure of Matter and the Solid State

energy and the single valence electron in the L shell with a higher energy. The lithium atom can be excited (as in a discharge in lithium vapour) so the valence electron can be given temporarily still higher energy to occupy various discrete levels, normally unoccupied like that shown in Figure 1.8(a).

In solid lithium (actually a body-centre cubic crystal form), however, the sharp energy levels broaden into bands as shown in Figure 1.8(b). This broadening is due to the proximity of neighbouring atoms. Whereas in the isolated atom the electrons are under the electrostatic influence of only the nucleus of the atom about which they circulate, in the solid state the electrons are also under the electrostatic influences of the nuclei of the immediately neighbouring atomic nuclei. Thus, the two valence electrons in a pair of neighbouring lithium atoms (considering the effect of only two very close atoms, for simplicity) may now occupy two energy levels, one slightly higher than that of the isolated atom and the other slightly lower. Whereas the single energy level in the isolated atom is E, there are now two energy levels at E_1 and E_2 spaced on either side of E. The value of $(E_1 - E_2)$ will increase the nearer the atoms are together.

Solid lithium has some 10^{28} atoms per cubic metre. There is the same number of valence electrons and twice this number of K shell electrons. Consequently there is an enormous number of energy levels which the electrons may ocupy within certain upper and lower limits, depending on the proximity of packing of the atoms in the crystal. These innumerable energy levels form a continuous band of energies.

For lithium the proximity effect on the inner K shell electrons is less than for the valence electrons in the outer L shell. The lower energy band occupied by K shell electrons is therefore narrower than the higher energy band occupied by the L shell electrons. Again a still higher unoccupied band exists which electrons can attain if the lithium atoms are excited (Figure 1.8(b)).

Hence, in N atoms of lithium, where N is a very large number even for a small specimen, the set of energy levels is not the same for every atom (as it is in the gaseous state) but is different for every atom. There are, indeed, N possible energy states in the energy band. It can be shown that two electrons (with opposite spins about their own axes) may occupy the same energy level but not more than two. N energy levels in a band are thus filled when they contain $2N$ electrons.

In the case of lithium, the lower energy state (derived from the K shell electrons) is filled; the valence band, however, is only half filled because it can contain $2N$ electrons in its N levels whereas there are only N valence electrons within N lithium atoms.

If an electric field is applied to a lithium specimen, the valence electrons will accelerate and so have their energy increased, but only if there are energy levels to which they can be raised. Since half the levels in the band are available, electrons can gain this extra energy, i.e. electrons can move: conduction occurs. The same concepts apply to all the monovalent metallic conductors, in particular, sodium and potassium.

There are, of course, several other metallic conductors which are divalent, for example, beryllium, magnesium and calcium. A

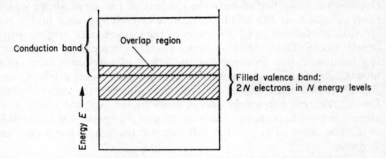

Figure 1.9. Energy band structure for a divalent metallic element where the conduction band overlaps the valence band

divalent metallic element will have two valence electrons per atom so the valence band will be filled. At first sight it would appear that energy cannot be given to these electrons on the application of an electric field. This is not the case because it turns out that these electrons can be excited into a conduction band which overlaps the valence band (Figure 1.9) and most of this conduction band (beyond the overlap region) is normally unfilled. The divalent metallic elements are therefore conductors because electrons can readily gain energy on the application of an electric field.

The energy band structures of all the metallic elements are such that conduction occurs because either the valence band is unfilled or the conduction and valence bands overlap.

1.17 Insulators

With the exception of sulphur and carbon in the form of diamond, the commonly encountered electrical insulators are compounds. A good example of an excellent insulator is aluminium oxide, Al_2O_3, two positive ions of aluminium are strongly bound by the three valence electrons per atom with three negative ions of divalent oxygen. The chemical bond energies are large, the valence electrons are firmly bound in the compound. None (or extremely few) are available for conduction in a specimen of alumina (the alternative name for aluminium oxide).

It is not an easy matter to deduce the energy band structures of insulators. However, they all have an important common feature. The valence band is filled and the higher unoccupied band into which the electrons could go is widely separated from the valence band (Figure 1.10). There is an *energy gap* of E_g electron-volt between the

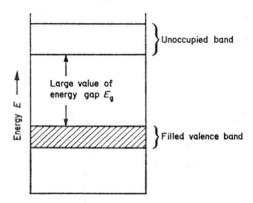

Figure 1.10. Typical energy band structure of an insulator

top of the valence band and the bottom of the nearest unoccupied level; E_g can typically be 9 eV.

For the purposes of rough calculation, an electron at a temperature T K can be said to have a most probable energy of kT, where k is the Boltzmann constant. In fact the electrons will have a distribution of energies, but most of them will have energies of about kT. A very small fraction of the total will have energies of $10kT$ and a much smaller fraction still will have energies of $100kT$.

The Boltzmann constant $k = (1/11\ 600)$eV (see Example 1.5(b)). In order to acquire an energy of 9 eV, the temperature would have to be

$$T = 9 \times 11\ 600 = 100\ 000 \text{ K} \quad \text{approx.}$$

This is meaningless because the insulator would vaporize. At only 1000 K, a very tiny fraction of the electrons would achieve energies of 9 eV, but so few that conduction would still be very small.

Note, however, that the conductivity of poor conductors (even though very small) increases with temperature rise, so insulators have a negative temperature coefficient of resistance.

1.18 Energy Band Structures of Semiconductors

As might be expected the energy band structures of the intrinsic semiconducting elements germanium and silicon are rather like those of insulators except that the energy gap between the valence and conduction bands is smaller.

To remove electrons from the covalent bonds of an intrinsic semiconductor (to form an electron-hole pair) requires a significant amount of energy. The energy gap between the top of the valence band and the bottom of the conduction band is 0.75 eV for germanium (Figure 1.11) and 1.1 eV for silicon.

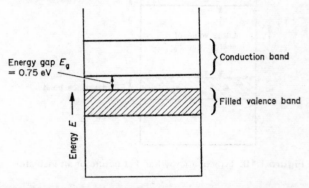

Figure 1.11. The valence and conduction band diagram for a germanium crystal

When a controlled amount of impurity is introduced into the lattice structure of an intrinsic semiconductor it can become an extrinsic semiconductor. Introducing a pentavalent impurity (to produce n-type germanium, for example) creates additional electrons to those in the covalent bonding and so creates new donor energy levels within the forbidden gap just below the conduction band (Figure 1.12(a)). The gap between this new level and the bottom of the conduction band is much less than the energy gap in the intrinsic semiconductor. For germanium it is 0.75 eV; when doped with pentavalent antimony, for example, the new level is only, perhaps, 0.05 eV or less below the bottom of the conduction band. Doped germanium is consequently a much superior conductor (with electrons as majority carriers) to pure germanium. Introducing a trivalent impurity (to produce p-type silicon, for example) creates additional positive holes and so produces new acceptor energy levels in the forbidden gap just above the top of the valence band (Figure 1.12(b)). Electrons from the valence band can readily be thermally excited into these vacancies leaving correspondingly vacant energy levels within the valence band which becomes partly unfilled. The conduction therefore increases greatly, with positive holes as the majority carriers.

Utilizing again a rough basis for calculation that an electron at a temperature T K has a most probable energy of $T/11\,600$ eV, it is seen that at absolute zero (0 K) semiconductors will be insulators. As the temperature is increased, conductivity increases. For an

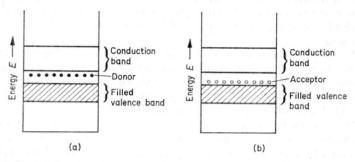

Figure 1.12. (a) Energy band diagram for a semiconductor doped with a pentavalent impurity and (b) energy band diagram for a semiconductor doped with a trivalent impurity

intrinsic semiconductor such as germanium, the fraction of the electrons at $T = 300$ K which gain enough energy to surmount the gap of 0.75 eV is sufficient for conductivity to occur. When this gap is reduced to only 0.05 eV by the controlled addition of a suitable pentavalent impurity element, it is easily seen that, at a given temperature, the fraction of the electrons able to enter the conduction band is increased greatly. Furthermore, it is apparent that conductivity increases with temperature so that the temperature coefficients of resistance of these intrinsic and extrinsic semiconductors are negative.

1.19 An Experiment to Determine the Energy Gap E_g for Germanium

For intrinsic germanium the equation which connects the width of the forbidden band, that is the energy gap E_g, with the conductivity σ is

$$\sigma = B \exp(-E_g/2kT) \qquad (1.38)$$

where B is a constant having the dimensions of conductivity, T is the absolute temperature and k is the Boltzmann constant. The theoretical derivation of this equation is based on the energy band theory of solids. An experimental verification of this equation consequently gives support to the concept of energy bands. Moreover, it enables the value of E_g for germanium to be determined.

Equation (1.38) can clearly be expressed in the alternative form

$$\log_e \sigma = \log_e B - E_g/2kT \qquad (1.39)$$

If, therefore, the resistance of a uniform block of germanium of known dimensions is determined at various measured temperatures T, σ is easily calculated and a plot of $\log_e \sigma$ against $1/T$ yields a straight line graph. The slope of this linear plot is $-E_g/kT$ so E_g is found.

A suitable experiment utilizes a short uniform rod of germanium about 15 mm in length, width 2 mm and thickness 2 mm. These dimensions are determined by a vernier or micrometer gauge. A constant current of 5 mA, say, is passed through this rod, which is provided with ohmic contacts at its ends.

This germanium rod is immersed in a beaker containing oil with the leads to the end contacts protruding above the surface. The oil is heated to various recorded temperatures (a mercury thermometer is used) by placing it on a gauze on a tripod beneath which is a Bunsen burner or suitable electrical heater. Maize oil obtainable from most grocers is suitable. Stirring of the oil is necessary to ensure a uniform temperature distribution.

At a number of recorded temperatures T within the range from about 20°C to 220°C, the potential difference V across the specimen is measured with a calibrated potentiometer or a high resistance voltmeter.

Providing that the current through the germanium rod is kept constant, σ is inversely proportional to the voltage V. Equation (1.39) therefore gives

$$\log_{10} V - \log_{10} C = 0.4343\, E_g/2kT - \log_{10} B$$

where C is another constant. A graph of $\log_{10} V$ against $1/T$ therefore gives a straight line of slope $0.4343 E_g/2k$, as in Figure 1.14.

It is convenient to use a constant current source for the supply of about 5 mA, otherwise frequent adjustments of a series resistance in a simple battery supply circuit will be necessary as the temperature is altered. The circuit diagram of a suitable current source (also useful in other experiments) is described in Section 4.14.

It is of interest to repeat this experiment with a doped specimen of germanium, e.g. n-type germanium containing antimony. Figure 1.13 shows the resistance plotted against the temperature for both an intrinsic and an extrinsic specimen.

The initial portion of the extrinsic curve indicates a small positive temperature coefficient of resistance similar to that for a metal.

In this temperature range up to about 300 K, all the impurity atoms are ionized (i.e. have lost an electron) and the number of electron-hole pairs produced does not compensate for the decrease in the

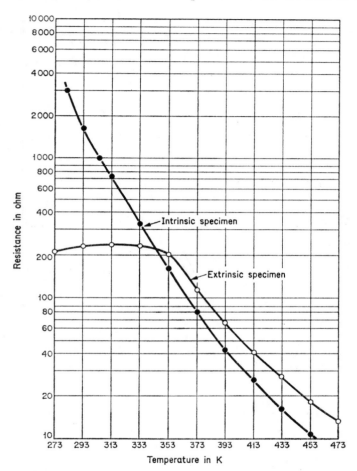

Figure 1.13. Plots of resistance against temperature for an intrinsic and an extrinsic specimen of germanium

mobility of the current carriers as the temperature rises. Hence the resistance of the specimen increases. At about 330 K, the intrinsic conductivity increases rapidly with temperature and so more than compensates for the decrease in mobility. At temperatures about 350 K, the resistance change with temperature is similar to that for

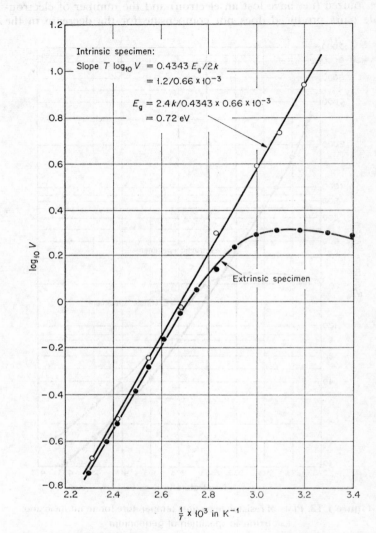

Figure 1.14. Plots of log V against $1/T$ for an intrinsic and an extrinsic specimen of germanium

The Structure of Matter and the Solid State

an intrinsic specimen. Figure 1.14 shows plots of $\log_{10} V$ against $1/T$ for the two specimens each of which yield a value of 0.72 eV for E_g.

Example 1.19(a)

Germanium contains 4.4×10^{28} atoms per cubic metre. Pentavalent antimony atoms are introduced into the lattice to a concentration of 1 part in 10^6. Calculate the resistivity of the extrinsic n-type germanium formed assuming that all the impurity atoms are ionized and the electron mobility is $0.36 \ m^2 \ V^{-1} \ s^{-1}$.

Conductivity $\sigma = nek_n$, where the majority carriers are electrons and where n = electron density = $4.4 \times 10^{28}/10^6 \ m^{-3}$, e = electronic charge = 1.6×10^{-19}C and k_n is the electron mobility = $0.36 \ m^2 \ V^{-1} \ s^{-1}$. Hence

Resistivity $\rho = 1/\sigma = 1/nek_n$

$$= \frac{1}{4.4 \times 10^{22} \times 1.6 \times 10^{-19} \times 0.36}$$

$$= 3.9 \times 10^{-4} \ \Omega \ m.$$

Example 1.19(b)

The resistivity of intrinsic germanium at 300 K is $0.47 \ \Omega \ m$. If the electron and positive hole mobilities are $0.37 \ m^2 \ V^{-1} \ s^{-1}$ and $0.16 \ m^2 \ V^{-1} \ s^{-1}$ respectively, calculate the density of the current carriers.

In an intrinsic material the electron and hole densities are equal, so

$$n_n = n_p = n_i$$

The conductivity $\sigma = J/E = n_i e (k_n + k_p)$ (see equation (1.37)), and the resistivity $\rho = 0.47 = 1/\sigma$. Hence

$$n_i = \frac{\sigma}{e(k_p + k_n)} = \frac{1}{0.47 \times 1.6 \times 10^{-19}(0.37 + 0.16)}$$

on putting the electronic charge $e = 1.6 \times 10^{-19}$C. Therefore

$$n_i = 2.5 \times 10^{19} m^{-3}.$$

Example 1.19(c)

With a constant current of 3mA passing through an intrinsic sample of germanium in the form of a uniform rod, the potential difference V across the specimen was recorded at a number of temperatures. The results are listed in the table given. Plot a graph of resistance against the absolute temperature T and also a graph of log V against $1/T$. Deduce a value for E_g, the width of the forbidden energy gap in electron-volt.

Temperature T K	$1/T$	V(mV)	$\log_{10}V$ (V in volt)
543	1.842×10^{-3}	13.4	$\bar{2}.1271$
532	1.883×10^{-3}	15.8	$\bar{2}.1987$
523	1.912×10^{-3}	17.8	$\bar{2}.2504$
512	1.952×10^{-3}	20.5	$\bar{2}.3324$
503	1.988×10^{-3}	25	$\bar{2}.3979$
493	2.028×10^{-3}	30	$\bar{2}.4771$
483	2.070×10^{-3}	36	$\bar{2}.5563$
473	2.114×10^{-3}	44.5	$\bar{2}.6484$
463	2.160×10^{-3}	53	$\bar{2}.7243$
453	2.208×10^{-3}	65	$\bar{2}.8129$
443	2.257×10^{-3}	79.5	$\bar{2}.9009$
433	2.309×10^{-3}	99	$\bar{2}.9952$
423	2.364×10^{-3}	126	$\bar{1}.1004$
413	2.421×10^{-3}	162.5	$\bar{1}.2108$
403	2.481×10^{-3}	205	$\bar{1}.3118$
383	2.611×10^{-3}	339	$\bar{1}.5302$
373	2.681×10^{-3}	449	$\bar{1}.6522$
363	2.755×10^{-3}	578	$\bar{1}.7619$
353	2.833×10^{-3}	762	$\bar{1}.8820$
339	2.950×10^{-3}	1046	0.0195
333	3.003×10^{-3}	1148	0.0599
323	3.096×10^{-3}	1236	0.0920
313	3.195×10^{-3}	1248	0.0962
303	3.300×10^{-3}	1211	0.0901
293	3.413×10^{-3}	1193	0.0766

The values of $1/T$ and $\log_{10}V$ required are included in the table in columns against the given values for T and V. The graph plotted is of the form shown in Figure 1.14 for intrinsic germanium. The value of E_g deduced from the slope of the graph as explained in section 1.19 is 0.72 eV.

1.20 The Hall Effect

If a metallic or a semiconducting specimen carrying a current I is placed in a region in which the uniform magnetic flux density is B where the flux lines are directed perpendicularly to the current flow (Figure 1.15(a)), the current carriers within the specimen are deviated so that a voltage is produced across the specimen in the direction perpendicular to both B and I. This *transverse galvomagnetic effect* was first noted by E. H. Hall in 1879 and is known as the Hall effect.

In addition to the appearance of the Hall voltage V_H across the specimen, the deviation of the current carriers in the magnetic flux will cause a small change in the resistance between the faces L and M. This *transverse magnetoresistance* is related to the Hall effect and can be detected as a small current change when the magnetic flux density B is changed.

It is convenient in an experiment on the Hall effect to use the constant current source described in Section 4.14 because frequent adjustments of the current through the specimen are avoided.

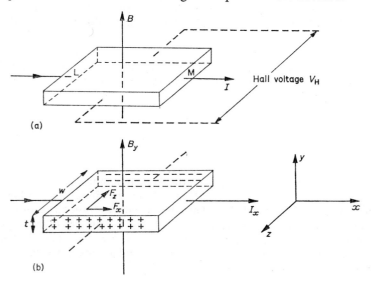

Figure 1.15. The Hall effect

The Hall effect is much larger in semiconductors than in metals and is therefore easier to measure. It enables the majority carriers to be identified as either electrons or holes and also their concentration to be determined. Hence the material can be quickly identified as n-type or p-type and the impurity concentrations can easily be calculated.

The magnitude of the effect can be readily calculated for the motion of current carriers of charge q (which would be $-e$ for an electron and $+e$ for a positive hole).

Referring to Cartesian axes Ox, Oy, Oz (Figure 1.15(b)) let the current through the uniform specimen in the direction Ox be I_x. The current per unit cross-section area, i.e. the current density. J_x, is therefore given by

$$J_x = I_x/wt \qquad (1.40)$$

where w is the width and t the thickness of the specimen cross-section.

If E_x is the corresponding electric field strength across the specimen in the x-direction, n is the carrier concentration (number of current carriers per unit volume) and k is the mobility of the current carriers,

$$J_x = nqkE_x \qquad (1.41)$$

The average force on a moving charge q due to the magnetic flux is in the z-direction (Fleming's left-hand rule) and is of magnitude

$$F_z = B_y qu$$

where B_y is the component of the magnetic flux density in the y-direction and u is the drift velocity in the x-direction. From equation (1.26), $J_x = nqu$. Hence

$$F_z = B_y q J_x / nq$$

Using equation (1.41),

$$F_z = B_y qkE_x \qquad (1.42)$$

The deviation of charge by the magnetic flux will cause free charge to accumulate on the faces of the specimen perpendicular to the z-axis (Figure 1.15(b)). This charge will continue to accumulate until an equilibrium state is established. At this equilibrium, the force qE_z on the charge q due to the growing electric field E_z (produced by the free charge) will be equal and opposite to that force produced by the magnetic flux. Therefore

$$qE_z = F_z = qkE_x B_y$$

from equation (1.42). Therefore

$$E_z = kE_x B_y$$

Substituting for kE_x from equation (1.41),

$$E_z = J_x B_y / nq \qquad (1.43)$$

The *Hall coefficient* R_H is defined by

$$R_H = E_z / J_x B_y$$

Hence, from equation (1.43)

$$R_H = 1/nq \qquad (1.44)$$

For a semiconductor, a more advanced theory gives

$$R_H = 3\pi/8nq \qquad (1.45)$$

As n is the carrier concentration in number per cubic metre, and q, the charge, is in coulomb, the unit of R_H is seen from equation (1.44)

The Structure of Matter and the Solid State

or (1.45) to be metre cubed per coulomb (m³ C⁻¹). Normally, the Hall voltage due to the electric field component E_z is measured. This is given by

$$V_H = E_z w \tag{1.46}$$

Substituting into this equation (1.46) the value of E_z given by equation (1.43),

$$V_H = \frac{J_x B_y w}{nq}$$

which, from equation (1.40), becomes

$$V_H = \frac{I_x B_y}{tnq} \tag{1.47}$$

For a semiconductor (equation 1.45) this becomes

$$V_H = \frac{3\pi I_x B_y}{8tnq} \tag{1.48}$$

This Hall voltage is measured between metal probes attached to the opposite faces of the crystal specimen.

Thin wafers of n-type germanium* (dimensions 5 × 5 × 0.4 mm approx.) are suitable for Hall effect measurements. It is not easy, however, to make good electrical contacts on the crystal slice. A preferable experiment is therefore to use a Hall probe, which is a single crystal slice with four leads attached (Figure 1.16(a)). Then, instead of undertaking actual Hall effect measurements, one can establish experimentally that, in accordance with equation (1.48), V_H is proportional to I_x and also to B_y. Having calibrated the probe, a constant current is passed through from the type of source described in Section 4.14, and it is used as a direct reading fluxmeter, making use of equation (1.48).

The four leads attached to the crystal slice (Figure 1.16(a)) are A and B to carry current and C and D for Hall voltage measurements. When a current flows between A and B but no magnetic flux is applied, the voltage between C and D should be zero. For this to be the case, C and D must be located on an equipotential. It is not possible to arrange this. Consequently, the arrangement show in Figure 1.16(b) is used. This provides a virtual contact which is often called a *virtual probe*. With current flowing through the probe but no applied magnetic flux, the contact RV1 is adjusted until the voltage between E and D is zero. The magnetic flux is then applied and the Hall voltage V_H is measured between E and D.

* Obtainable from *Mullard Educational Service*, who also publish a leaflet on Hall effect experiments.

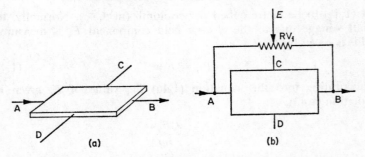

Figure 1.16. (a) A typical Hall probe and (b) use of a virtual contact to 'zero' the Hall probe

The identification of majority carriers by means of the Hall effect is possible. Suppose the current I (conventional direction from $+$ to $-$) flows through a specimen (Figure 1.17). If the majority carriers are electrons, they will move towards the left in Figure 1.17(a) and the force F on them upwards due to the transverse magnetic flux is of density B (Fleming's left-hand rule). Negative charge therefore accumulates on the top surface (contact C) and positive on the bottom surface (contact D). The voltage across CD is therefore with

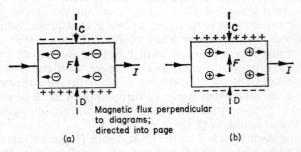

Figure 1.17. The identification of majority carriers by means of the Hall effect

a certain polarity, D being positive with respect to C. When the majority carriers are positive holes, the direction of the force on them due to the magnetic flux is still the same. Now, however, the top surface becomes positively charged and the bottom surface negatively charged: the polarity of the voltage is reversed (Figure 1.17(b)).

Suppose an air-cored coil carries a current I_B. The magnetic flux density B_y produced within the coil is proportional to I_B. The Hall voltage V_H established across a semiconductor specimen within this

The Structure of Matter and the Solid State

flux is proportional both to the current I_x through this specimen and to B_y, i.e. to I_B. Hence

$$V_H \propto I_x I_B$$

A device of value in computers is thereby indicated in principle: it is capable of multiplying two quantities of magnitudes decided by I_x and I_B and where the product is proportional to V_H.

Example 1.20(a)

A copper strip carrying a current of 3 A is placed in a uniform magnetic flux of density 1.5 tesla with the flux lines at right angles to the direction of the current flow. If the width of the strip is 10.0 mm and its thickness is 0.15 mm, calculate the Hall voltage that appears across the strip if the number of copper atoms per cubic metre is 9×10^{28}. Assume that each atom contributes one free electron. (The electronic charge $e = 1.6 \times 10^{-19}C$).

From Equation (1.47)

$$V_H = \frac{I_x B_y}{tnq} = \frac{3 \times 1.5}{1.5 \times 10^{-4} \times 9 \times 10^{28} \times 1.6 \times 10^{-19}}$$

$$= \frac{4.5}{21.6} \times 10^{-5} = 2.1 \mu V.$$

Note that this voltage is very small even though the strip is carrying a current of 30 A in a high magnetic flux density. As a result Hall voltage measurements on metals are not usually attempted in school or college laboratories.

Example 1.20(b)

A rod of n-type germanium is 2.0 mm wide and 0.15 mm thick. A current of 10.0 mA is passed along its length and a uniform magnetic flux of density 0.2 tesla is established at right angles to the current flow. The Hall voltage is 2.5 mV. Calculate the Hall coefficient and the electron concentration.

For a semiconductor equation (1.48) applies:

$$V_H = \frac{3\pi I_x B_y}{8tnq}$$

The electron concentration is therefore

$$n = \frac{3\pi I_x B_y}{8 V_H et}$$

because $q = e$, the electronic charge. Substituting the values given together with $e = 1.6 \times 10^{-19}C$,

$$n = \frac{3\pi \times 10^{-2} \times 0.2}{8 \times 2.5 \times 10^{-3} \times 1.6 \times 10^{-19} \times 1.5 \times 10^{-4}} m^{-3}$$

$$= 3.9 \times 10^{22} m^{-3}$$

From equation (1.45),

$$R_H = \frac{3\pi}{8nq} = \frac{3\pi}{8 \times 3.9 \times 10^{22} \times 1.6 \times 10^{-19}} \text{ m}^3\text{ C}^{-1}$$
$$= 1.9 \times 10^{-2} \text{ m}^3\text{ C}^{-1}.$$

Notice the comparatively large Hall voltage V_H quoted for this n-type germanium sample even though a modest current and magnetic flux density are used. A sample such as this one is ideal for simple experimental work.

1.21 Identification of n- or p-type Germanium by means of a Cooled Probe

A very simple method of identifying whether a germanium sample is n- or p-type without making use of the Hall effect is to use a cooled probe. Two copper wires are attached to a centre zero millivoltmeter. The free ends of the wires are bare and held on rods of insulating material (Figure 1.18).

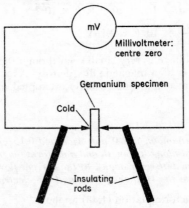

Figure 1.18. Using a cooled probe to determine whether a germanium sample is n- or p-type

Assume that when the right-hand lead is positive with respect to the left-hand lead the deflection is to the right and *vice versa*. The free tip (probe) of the left-hand lead is cooled by holding it in contact with ice or solid carbon dioxide. The two probes are placed on either side of the germanium slice. If the deflection recorded by the millivoltmeter is to the right, the cooled probe is negative so the germanium is n-type. A deflection to the left indicates that the germanium is p-type. The generated e.m.f. is thermoelectric in origin; the thermoelectric e.m.f. from a semiconductor is larger than that from a metal.

The Structure of Matter and the Solid State

In the following exercises the abbreviation A.E.B. denotes that the question which it terminates has been taken with permission from papers set in Advanced level examinations by the Associated Examining Board, either in Physics or in the endorsement paper on Electronics.

Exercise 1

1. What information regarding the forces between molecules or atoms can be inferred from the physical properties of materials in the solid, liquid and gaseous states?
   Explain in terms of molecular motion why liquids cool as they evaporate.
2. Explain the terms *amorphous, crystalline, polycrystalline* and *isotropic*.
3. Calculate the speed of an electron which has been accelerated from rest through a potential difference of 3500 V. (Specific charge of electron, e/m_e, is 1.76×10^{11} C kg^{-1}).
4. Calculate the root mean square speed of molecules of oxygen in a container at a temperature of 300 K and a pressure of 1.013×10^5 N m^{-2}, given that the density of oxygen at s.t.p. is 1.62 kg m^{-3}.
   An oxygen molecule is allowed to fall freely through a distance of 1.0 m. Calculate the energy acquired by this molecule and express this as a fraction of the average kinetic energy of an oxygen molecule at 300 K.
   (Molecular mass of oxygen = 32.00; Avogadro constant = 6.023×10^{26} per kilomole; gravitational acceleration = 9.8 m s^{-2}; Boltzmann constant = 1.38×10^{-23} J K^{-1}).
5. Electrons which have been accelerated from rest through a potential difference of 2.0 kV are injected into a uniform magnetic flux of density 2×10^{-3} tesla in a direction at right angles to the flux lines. Calculate the radius of curvature of the electron trajectory. (Specific charge of electron, e/m_e, is 1.76×10^{11} C kg^{-1}).
6. 'Free electrons in a metallic conductor behave in a similar fashion to the molecules of an ideal gas'.
   Discuss the behaviour of electrons in metals which supports this statement.
7. Outline aspects of the behaviour of the elements sodium, carbon and silicon that can be inferred from the knowledge that their atomic numbers are 11, 6 and 14 respectively.
8. Compare and contrast the electrical behaviour of metals and semiconductors.
9. Explain the classification of materials into conductors, insulators and semiconductors in terms of electron energy levels. (A.E.B.)
10. The atomic weight of sodium (Na) is 22.99 and that of chlorine (Cl) is 35.46. If the density of a large crystal of rock salt (NaCl) is 2.163×10^3 kg m^{-3}, calculate its lattice constant (the separation between neighbouring parallel planes of atoms. (Avogadro constant = 6.025×10^{26} per kilomole).
11. Discuss the electronic structure of atoms of elements which may be used to dope a specimen of germanium to produce (*a*) n-type germanium and (*b*) p-type germanium.

How does the process of doping modify the energy band structure of a semiconductor?
12. Write short notes on
    (i) intrinsic and extrinsic conductivity;
    (ii) majority and minority carriers;
    (iii) carrier mobility. (A.E.B.)
13. What is meant by (*a*) intrinsic, and (*b*) extrinsic conductivity?
    Explain how and why the electrical conductivity of a crystal slice of germanium varies with:
    (*a*) temperature over the range 273–523 K;
    (*b*) trivalent impurity atoms present;
    (*c*) pentavalent impurity atoms present. (A.E.B.)
14. The following two experiments are undertaken:
    (*a*) the resistance of a rod of intrinsic germanium is measured as its temperature is raised from 300 K to 500 K;
    (*b*) the resistance of a rod of n-type germanium is measured as its temperature is raised from 200 K to 500 K.
    Sketch for both cases the curve of resistance against temperature which would be obtained and explain the processes which determine the shapes of these curves.
    How could the energy gap (the width of the forbidden band) for germanium be determined from either set of these experimental results?
15. Derive an expression for the conductivity of a semiconducting material in terms of the concentration n and the mobility k of the electrons and holes.
    A rod of instrinsic germanium of dimensions 10.0 mm × 2.0 mm × 1.0 mm has a resistance of 150 Ω at a temperature of 400 K. If the electron and hole mobilities are respectively 0.37 m^2 V^{-1} s^{-1} and 0.18 m^2 V^{-1} s^{-1}, calculate the electron concentration in the specimen at this temperature. (Electronic charge $e = 1.6 \times 10^{-19}$C.)
16. An intrinsic specimen of germanium at 300 K has a resistivity of 0.47 Ω m. Calculate the concentration of electrons if the electron and hole mobilities are respectively 0.36 m^2 V^{-1} s^{-1} and 0.17 m^2 V^{-1} s^{-1}.
17. For germanium at a certain temperature it is known that the number of electrons per unit volume and of holes taking part in intrinsic conductivity is 3.0×10^{19} m^{-3}. At this same temperature, a rod of p-type germanium of dimensions 10 mm × 2 mm × 0.5 mm has a resistance of 100 Ω. Calculate the impurity concentration in this specimen assuming that all the acceptor atoms are ionized. (Electron charge $e = 1.6 \times 10^{-19}$C; electron mobility $k_n = 0.39$ m^2 V^{-1} s^{-1}; hole mobility $k_p = 0.19$ m^2 V^{-1} s^{-1}).
18. What is the Hall effect? Explain how Hall voltage measurements on a semiconducting crystal slice can be used to identify the majority carriers and to determine their concentration.

2 The manufacture of semiconductor devices

Although a very large number of semiconducting materials is known, only a relatively small number is at present being used to make electronic components. Germanium and silicon in single-crystal form are by far the most common elements used but the search continues in laboratories throughout the world for other suitable materials. Whether in single-crystal or polycrystalline form, the aim is to provide cheaper, more reliable and more versatile semiconductor devices.

In the past decade silicon has replaced germanium in most components and this preference for silicon continues to grow. Germanium was used first to make most of the transistors because it is easier to purify. Subsequently, silicon has been proved to have superior characteristics. In particular, a silicon component can operate over the wide temperature range from $-50°C$ to $150°C$ with negligible deterioration in its electrical behaviour.

The problem which has faced the semiconductor industry appears simple: to produce a junction between p- and n-type materials to make a rectifier, and two such junctions in a single component to make a junction transistor. The solution has nevertheless proved very difficult and has demanded an almost unprecedented effort to provide semiconducting elements of adequate purity with controlled additions of different known elements, to manufacture junction transistors with exceedingly close p-n junctions, and provide known crystal structures with reliable and reproducible electrical characteristics.

In this unique industrial development chemists contribute by providing methods of extracting and purifying the elements and investigating reactions which clean, etch and stabilize the crystal surface. Physicists explain on a basis of solid-state physics the behaviour of existing components and provide, in conjunction with electronic engineers, interesting new designs and circuit applications. Metallurgists study the production of alloys and the techniques of growing large single crystals while mechanical engineers develop intricate machines for slicing, handling and encapsulating the miniature components. Often the environment under which these semiconductor devices are assembled is more akin to a hospital operating theatre than a factory. Indeed, it is difficult to convey the fantastic effort that has been necessary to enable a reliable transistor to be purchased for at low cost.

2.1 The Zone Refining of Germanium and Silicon

Germanium is in group IVA of the periodic table which comprises carbon, silicon, germanium, tin and lead, all quadravalent elements of which the atomic numbers are respectively 6, 14, 32, 50 and 82. Germanium is a rare element with a grey metallic appearance and is generally extracted from sulphide ores of zinc, lead and copper in which it occurs in low concentration. It is also found in the ashes of certain coals.

Silicon is the second most abundant element in the earth's surface

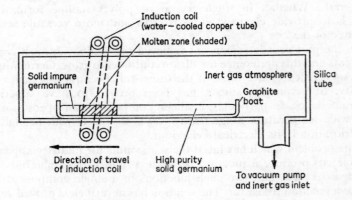

Figure 2.1. The technique of zone-refining

but usually occurs in combination with oxygen as silica (SiO_2) and is difficult to prepare in a pure state.

For the semiconductor industry, both silicon and germanium are obtained in a pure state from the tetrachloride or the dioxide. For the manufacture of semiconductor diodes and transistors, however, the demands on purity are considerably greater than is usually encountered in the most refined analytical chemistry. Whereas an impurity concentration of 1 part in 10^6 is considered top class work in normal chemical practice, only about 1 foreign atom in 10^9 atoms can be tolerated in germanium or silicon suitable for semiconductor devices.

The Manufacture of Semiconductor Devices

The most powerful technique for obtaining the necessary super-purity is by the process of zone-refining. This method depends on the fact that the impurities are more soluble when the element is in the liquid state. Hence, if by heating a molten zone is made to traverse a bar of germanium, the impurities will tend to concentrate in the molten region and be swept to one end of the bar. A number of such passes will concentrate the impurities at one end which can be sawn off from the cooled bar.

Property	Germanium	Silicon
Appearance	Grey metallic lustre	Bluish-grey metallic lustre with blue hue due to oxide layer
Nature	Hard and brittle	Hard and brittle
Atomic number	32	14
Atomic mass	72.60	28.06
Valence	4	4
Density (kg m^{-3})	5.3×10^3	2.3×10^3
Melting point (K)	1230	1693
Crystal type	Diamond	Diamond
Approximate intrinsic resistivity at 300 K (Ω m)	0.50	2.3×10^3
Electron mobility (m^2 V^{-1} s^{-1})	0.39	0.13
Hole mobility (m^2 V^{-1} s^{-1})	0.19	0.05

Table 2.1
Properties of germanium and silicon

To undertake such zone refining a bar of the germanium (melting point: 958°C) is contained in a quartz or graphite boat placed in a silica tube which is either evacuated or filled with an inert gas such as argon to prevent contamination. The heating of a zone of the bar is arranged by an induction coil around the silica tube. This coil carries a radio-frequency alternating current of several amperes; the electromagnetic radiation it produces sets up eddy-currents in the germanium rod (Figure 2.1).

A similar refining process which eliminates the use of a container boat has been successfully applied to silicon which melts at 1420°C. In this *floating zone process* the bar of silicon is clamped in a vertical position in a quartz tube. Again, a molten zone is made to traverse the bar, the molten region being held in position by surface tension forces. Very careful temperature control is demanded. A doping material added at one end can be uniformly distributed along the bar by means of a single cycle of zone refining.

2.2 Growing Large Single Crystals

One method often used to grow large single crystals of germanium or silicon for the manufacture of transistors is due to Czochralski (Figure 2.2). Within an inert atmosphere an electrically heated

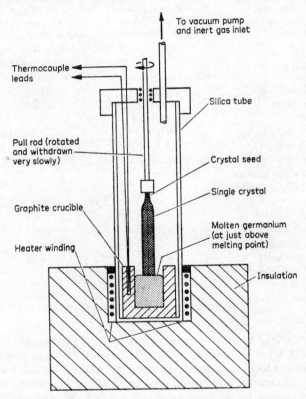

Figure 2.2. The Czochralski method of growing single crystals

crucible maintains the germanium at a temperature just a few degrees above its melting point. A very small piece of single crystal, called a *seed*, is lowered on to the surface of the melt and then slowly withdrawn. Surface tension forces support some of the molten material, which cools slowly and solidifies with an orderly arrangement of atoms identical with that of the seed. Over a period of perhaps twenty-four hours, a single crystal 30 mm in diameter and 300 mm long could be grown.

Trivalent or pentavalent doping elements can be added to the melt before the crystal is grown to produce a p- or n-type crystal of known resistivity. This resistivity, governed by the impurity content, will depend on the rate at which the impurities are transferred from the melt to the crystal and can be controlled by varying the crystal growth rate.

2.3 Forming a p-n Junction

A junction is ideally a surface separating a p-type semiconductor from an n-type. The very useful electrical properties of such a junction cannot be reliably obtained by placing a p-type material in contact with an n-type. Indeed, the transition must be produced within a single crystal slice.

In some components the transition needs to be as sharp as possible; in others, the transition needs to extend over a finite region of the crystal. These two extremes are referred to as *step* and *graded* junctions respectively. Of the many processes used to produce a sharp change in the impurity concentration within a semiconductor crystal lattice, three will be described in some detail.

The first is called *rate-growing*: it enables a p-n junction to be created merely by varying the rate at which the crystal is allowed to grow from the melt. In the second method, an *alloy* is formed of the semiconductor and the chosen impurity. *Diffusion* is the third method: the semiconductor is heated in a vapour of the impurity atoms which enter the crystal surface and diffuse slowly into the crystal lattice. The depth to which the impurity atoms penetrate is governed by the temperature and the time. This diffusion process normally takes several hours and temperatures in the range 500°C to 600°C are used for germanium and 900°C to 1300°C for silicon.

2.4 Rate Growing

Rate growing of a p-n junction in germanium was first reported in the early 1950s. The method is based on the fact that the equilibrium of the impurity (the doping material) between the solid and liquid phase depends on the rate at which the solid is formed, i.e. on the rate of crystal growth. It is possible to introduce both an acceptor and a donor impurity and by suitably adjusting their melt concentrations to incorporate an excess of acceptor in the crystal at one growth rate, to give a p-type semiconductor, and an excess of donor at a different growth rate, to yield an n-type semiconductor.

This principle has been extended to produce the n-p-n or a p-n-p structure required for transistors merely by suitable variation of the growth rate. Here n-p-n implies a p-type material sandwiched between n-type specimens, so two junctions are provided: n-p and p-n in order. Likewise, for p-n-p.

Although used in the manufacture of junction transistors for several years, this techniques has been largely replaced by the alloying and the diffusion processes.

2.5 Alloy Junctions

A large n-type crystal of germanium of resistivity about 5×10^{-2} Ω m (the impurity doping element is usually pentavalent antimony) is

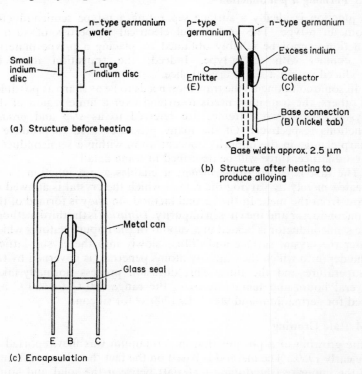

Figure 2.3. The alloying method and a p-n-p junction transistor

cut into slices each approximately 0.3 mm thick along a crystal plane which is identified by x-ray diffraction.

As they are very brittle, these slices are mounted in wax before being passed under diamond-impregnated cutting wheels to produce hundreds of tiny wafers of dimensions perhaps $3.0 \times 7.0 \times 0.3$ mm. A pellet of indium (trivalent) is placed on one face of each crystalline wafer. The wafers are then heated in a hydrogen atmosphere in a furnace. The indium melts (melting point 156°C) and dissolves some of the germanium from the wafer. The furnace temperature is carefully controlled during the heating and cooling cycle. As the molten alloy (germanium-indium) cools, the dissolved germanium recrystallizes at the liquid-solid interface, the wafer acting like a single crystal seed. The recrystallized germanium is p-type because it contains, in solid solution, some of the indium. A nickel tag is soldered to the germanium wafer and a wire to the pellet con-

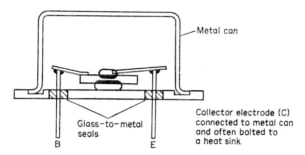

Figure 2.4. A power transistor

taining the excess indium. The device formed, consisting of a p-n junction, is a semiconductor junction diode. The assembly is etched to remove surface contamination, washed in de-ionized water, dried and mounted in water-repellent grease in a light-tight capsule.

Two junctions can be produced very close together by placing spheres or discs of the doping material (trivalent indium) on opposite sides of the n-type germanium wafer. Figure 2.3 illustrates the process and shows the final form of a p-n-p junction transistor made in this way.

The indium pellet used to form the collector electrode of the p-n-p transistor is always larger than that used to form the emitter. In a germanium power transistor, the materials and the process are the same except that materials used have larger dimensions, and the collector electrode is bonded to the metal housing (Figure 2.4). When mounted in a circuit the metal can of the power transistor is bolted to a heat sink which becomes the collector electrode.

2.6 Junctions Produced by Diffusion

The production of p-n junctions by diffusion is of considerable importance in transistor technology. Diffusion methods make it possible to control very precisely the concentration and the concentration gradient of an impurity over very small regions of a semiconductor. As an example of this process, the construction of a silicon planar diode is described.

The planar process comprises three parts:

(a) Covering a slice of n-type silicon with a hard inert oxide layer which acts as a diffusion mask.
(b) Etching holes in the oxide layer to expose and define those areas of silicon into which impurity may diffuse.
(c) Diffusing in from the vapour phase the trivalent impurity, which is usually boron.

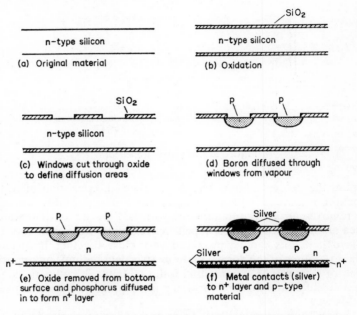

Figure 2.5. The diffusion process in making a silicon planar diode

Steps in the construction process are shown in Figure 2.5. From a slice of n-type silicon approximately 25 mm in diameter and 0.15 mm thick, about 1000 low-power diodes can be constructed. A layer of oxide (SiO_2) is formed on the slice by heating it to high temperatures in a steam atmosphere. Holes are cut in the oxide layer to expose the silicon. In another high temperature tubular furnace, boron is diffused from the vapour into the exposed areas of silicon to form a p-n junction. The oxide which covers the back surface of this slice is now removed and pentavalent phosphorus is diffused into the back in a third high temperature furnace. This

The Manufacture of Semiconductor Devices

very highly doped n region is normally termed an n^+ region, and is so labelled in Figure 2.5. This term can be misleading: it is merely used to signify that the region is doped very heavily, has a conductivity similar to that of a metal and allows an ohmic (non-rectifying) contact to be made to it without difficulty.

The centre of the p-type region is exposed and silver or gold contacts are made, one to the p-region and the other to the n^+ region.

The planar diode has an oxide covering except where the contacts are exposed. This inert layer eliminates the very troublesome electrical behaviour of a contaminated surface. The surface is said to be *passivated* and the diode is known as a *passivated planar diode*.

It is apparent that this diffusion process together with oxide masking can be employed to create junction transistors with well controlled characteristics. In addition, the technique allows a complex circuit to be constructed on a single slice of silicon. This circuit, called an *integrated circuit* (IC) when mounted may be only a little bigger than a single transistor. Great efforts in microcircuits of this kind are proving very fruitful and provide enormous savings in space and weight for space vehicle electronics and ever more complex computers.

Exercise 2

1. Explain, with the aid of a diagram, the zone-refining process used to obtain germanium in a very pure state.
2. A contact between a metal and a semiconductor is specified as an ohmic contact. Explain the meaning of this term and outline one method used to produce such a contact.
3. Describe, with a diagram of the apparatus, a method of growing large single crystals of germanium.
4. Describe, with diagrams, the process of forming a p-n junction (*a*) in germanium by alloying and (*b*) in silicon by diffusion.

3 Semiconductor diodes and rectification

3.1 Power Supply

The operation of electronic apparatus almost always requires power supplies. The familiar example is the radio valve (thermionic vacuum tube) which needs a low voltage (low tension, L.T.) supply for its filament or cathode and a high voltage (high tension, H.T.) supply across its anode and cathode. Semiconductor devices, such as the transistor, offer immediate practical advantages in this respect: there is no filament or cathode heater and the supplies needed are normally at much lower voltages than those required for thermionic vacuum tubes: in most cases only a few volts is required as against 100 volts or more.

Dry batteries such as the familiar six-cell 9 volt ones, are often used as power supplies for semiconductor electronic circuits, the well-known example being the transistor radio set; another instance is the electronic measuring instrument, for example, the transistor voltmeter.

There is, however, the need to provide a steady voltage supply (a so-called d.c. voltage) by making use of the a.c. mains. The a.c. mains supply is commonly at 240 V r.m.s. and alternates at a frequency of 50 Hz. This voltage is frequently too high or too low for the practical purpose required. It is therefore first changed to a different value by means of an a.c. transformer. This transformer is of the step-down type ($T < 1$ where T is total number of turns on the sscondary winding divided by the total number on the primary winding) if the output r.m.s. voltage is to be less than 240 V, and of the step-up variety ($T > 1$) if the output is to exceed 240 V r.m.s.

This transformer has the desirable feature also that the output is isolated, as regards direct electrical connection, from the mains.

The secondary voltage is alternating: its output varies sinusoidally with time (Figure 3.1(a)). It is essential to render this output uni-directional to provide eventually a d.c. voltage. This means that the positive half-cycles, say, are needed, but the negative half-cycles have to be eliminated. The process whereby this is achieved is called *rectification*. A device called a *rectifier* is needed which conducts electricity readily when the voltage across it is in one direction but which conducts electricity poorly (ideally, not at all) when the

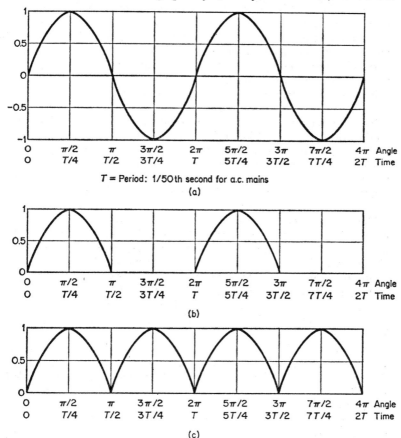

Figure 3.1. (a) The sine waveform of an a.c. mains supply, (b) half-wave rectification: the waveform, and (c) full-wave rectification: the waveform

voltage is reversed. The semiconductor diode is the modern and convenient rectifier.

Elimination of the half-cycles of one polarity (say the negative ones) results in a voltage of the wave-form shown in Figure 3.1(b). This is known as *half-wave rectification*; if, by circuit ingenuity, the undesired half-cycles are not eliminated but reversed in polarity, *full-wave rectification* (Figure 3.1(c)) is performed.

The half-wave or full-wave rectified supply is not satisfactory for most power supply purposes. It cannot be envisaged that the voltage across the electronic apparatus is allowed to vary between zero and a peak value each cycle or twice a cycle. Imagine the unfortunate noise that would result from a radio set with such a power supply!

It is therefore necessary to smooth these wide fluctuations of voltage. This means that the voltage has to be constant and not vary with time. The graph of voltage against time would then be simply a straight line parallel to the time axis. This smoothing is achieved by capacitors or, better, a smoothing filter involving capacitors and inductors or resistors.

Though steady d.c. voltage supplies are hence obtained by rectification and smoothing, it is frequently the case in the operation of electronic apparatus used for measurement purposes that the voltage is not sufficiently constant. Thus, it might be constant to within ± 1 per cent but the demand is that it be constant to within ± 0.001 per cent or better. This additional requirement is met by a *voltage stabilizing circuit*.

The most widely used rectifier is the silicon p-n junction, known as the *silicon diode*. The simplest stabilizer is a silicon p-n junction used as a *Zener diode*, also known as a *reference voltage diode*. Together, their use has revolutionized power-pack design.

Stabilization of current is also a requirement as well as stabilization of voltage. Zener diodes as reference voltage devices are described in this chapter. The more complex circuits capable of maintaining either the voltage or the current at a fixed value will be examined in Chapter 4.

Apart from their obvious advantages of size and weight compared with rectifier valves and the old metal rectifiers, silicon diodes have proved to be very efficient and reliable where high power is demanded as well as for the small power requirements of small-scale electronic apparatus.

In Figure 3.2(a) is shown a selection of semiconductor components; included is a silicon diode capable of passing a current of 20 A. Although the mounting and cooling of this high-power component is a separate problem from the concerns of the small power apparatus described in this text, yet its electrical characteristics are essentially similar to those of the modest diodes.

Transistor outlines (TO) have internationally agreed sizes. Two are T05 and T018 of which typical dimensions in mm are tabulated (Figure 3.2(b)):

	A	B	C	D	E	F	G	H	J
T05	9	8	6	5	0.8	40	0.5	0.4	0.9
T018	6	5	5	3	1	13	0.4	0.05	1

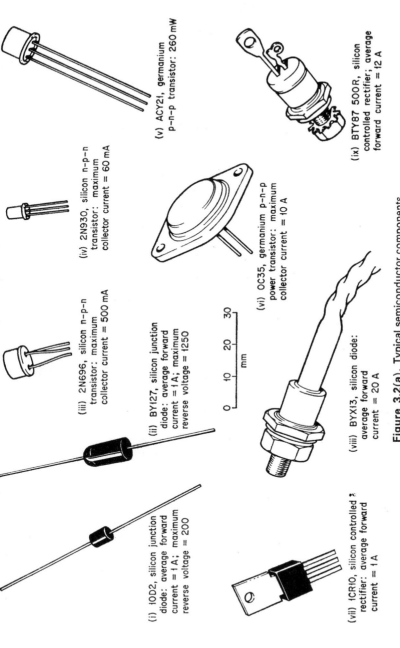

Figure 3.2(a). Typical semiconductor components

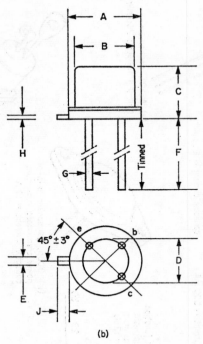

Figure 3.2(b)

3.2 The Electrical Characteristics of the p-n Junction Diode

The diode, a two-terminal device, offers negligible resistance to the flow of current in one direction and yet presents a nearly infinite resistance in the opposite direction. The circuit symbol (Figure 3.3(a)), which is often printed on the component case (Figure 3.3(b)), shows how the device should be connected for a particular behaviour.

The conventional current direction is from positive to negative (that in which a positive charge would move); the electron current is the other way, from negative to positive. Conventional current flows easily in the direction indicated by the arrow (Figure 3.3(a)).

When the terminal connected to the arrow head (the anode) is positive with respect to the other terminal, represented by a short line against the point of the arrow head (the cathode), the diode is said to be *forward biased*. When the diode is forward biased, conventional current flows readily in the direction of the arrow; electrons flow in the opposite direction from cathode to anode.

If the polarity of the voltage across the diode is reversed (arrow head negative; other terminal positive) the diode is *reverse biased*.

Then only a small leakage current of a few microamperes will flow.

Sometimes, instead of the circuit symbol being printed on the component casing, a band is marked around this casing at the cathode

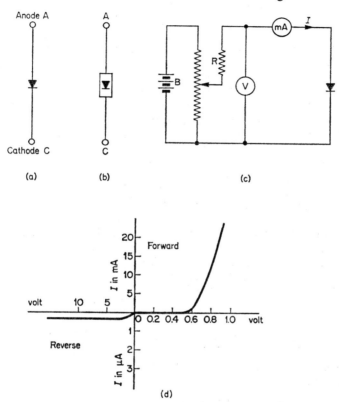

Figure 3.3. The silicon diode: (a) circuit symbol; (b) the component; (c) circuit for obtaining current against voltage characteristic; and (d) the current-voltage characteristic

end. Note that the terms 'anode' and 'cathode' are a carry-over from the days of thermionic vacuum diodes.

To determine the current through the silicon diode for various potential differences across it, the circuit shown in Figure 3.3(c) is used. The experiment with this circuit arrangement is in two parts. In part (i) the forward biased characteristic is obtained; in part (ii) the reverse biased.

For part (i) it is necessary to include a 50 Ω resistor R in series in the circuit to limit the forward current; otherwise the diode will be damaged if 1.5 V is placed directly across it. The maximum p.d. applied is 1.5 V and the moving-coil voltmeter V should be capable of reading 0 to 1.5 V in 0.1 V steps. The milliammeter mA has a full-scale deflection of perhaps 50 mA.

The current I recorded by mA is recorded for values of the p.d. V across the diode from zero in steps of 0.1 V up to 1.5 V. The plot of the forward characteristic is typically as shown in Figure 3.3(d).

To undertake part (ii) of the experiment, *the polarity of the battery B is reversed:* the arrow head (anode) is now made negative with respect to the cathode. The p.d. across the diode to obtain the reverse bias characteristic needs to be considerably larger than in part (i). The e.m.f. of the battery B is now about 10 V. The voltmeter V has to be changed for one with a scale reading from 0 to 10 V in 0.5 V steps (or the range of a multirange voltmeter is increased). The reverse current is only a fraction of a microampere corresponding to a reverse resistance of several megohm. The meter mA hence has to be changed to a microammeter capable of reading 0 to 0.1 μA, preferably in 0.01 μA steps: a lamp-and-scale box-type galvanometre of known current sensitivity is suitable.

Note that the vertical axis of Figure 3.3(d) is marked in mA in the positive direction and μA in the negative direction.

The current I which flows through a p-n junction at a temperature T K when a potential difference V is maintained across it is given quite accurately by the equation

$$I = I_0[\exp(eV/kT) - 1] \tag{3.1}$$

where I_0 is a constant current of value about 0.05 μA for a silicon p-n junction, k is the Boltzmann constant and e is the electronic charge. As $e = 1.6 \times 10^{-19}$ coulomb and $k = 1.38 \times 10^{-23}$ joule deg^{-1}K, so at $T = 300$ K (27°C)

$$\frac{e}{kT} = \frac{1.6 \times 10^{-19}}{1.38 \times 10^{-23} \times 300} \text{ coulomb per joule}$$
$$= 38.5 \text{ per volt} = 38.5 \text{ V}^{-1}$$

Equation (3.1) may therefore be written

$$I = I_0[\exp(38.5V) - 1] \tag{3.2}$$

where V is the numerical value of the p.d. in volt across the diode. This voltage is positive when the junction is forward biased and negative when it is reverse biased. With reverse bias, put $V = -1$ V in equation (3.2), then

$$I = I_0[\exp(-38.5) - 1]$$

Here $\exp(-38.5) = e^{-38.5} = 1/e^{38.5}$ is negligibly small compared with 1 because $e = 2.71828$. The current I is therefore $= I_0$.

Consequently, with a reverse bias exceeding about 1 V, the reverse current is constant at I_0, which is about 0.05 μA (Figure 3.3(d)). I_0 is known as the *reverse saturation current*. This reverse saturation or leakage current increases with temperature T. Nevertheless, silicon diodes can be operated with a performance only slightly below their optimum at temperatures up to 150°C.

In the forward direction, V is positive in equation (3.2). For the forward current to become a few milliampere, $\exp(38.5V)$ in equation (3.2) must become large compared with unity. With $I_0 = 0.05$ μA, the forward current I will be 1 mA approximately at $T = 300°C$ when

$$\exp(38.5V) = 1 \text{ mA}/0.05 \text{ μA} = 20\,000$$

Hence
$$38.5V = \log_e 20\,000$$
$$= 2.3 \log_{10} 20\,000$$
$$= 2.3 \times 4.3010 = 9.9$$
$$V = \frac{9.9}{38.5} = 0.26 \text{ V}$$

It follows that the current in the forward direction is very small until the forward bias exceeds 0.2 to 0.5 V; above these values, this current increases exponentially as shown in Figure 3.3(d), its value being limited only by the series resistance R. Indeed, if R is omitted, the diode will be ruined by a voltage of 1.5 V or so.

3.3 Explanation of the Electrical Behaviour of a p-n Junction Diode in Terms of its Structure

As soon as a junction is created between p-type silicon and n-type (indeed, for any p- and n-type materials), electrons which predominate in the n-type silicon will move across the junction to fill some of the preponderant positive holes (vacancies) in the p-type material. Conversely, holes move across the junction in the opposite direction from the p-type to the n-type.

On the n-type side of the junction, loss of electrons from the material will leave some atoms with an excess positive charge, i.e. positive ions are created. On the p-type side, negative ions are produced. These ions are locked in the crystal lattice structure: they are not mobile like the electrons and holes. An electric field (due to these ions) will therefore exist across the p-n junction and directed from the n-type to the p-type material (Figure 3.4(a)). The field direction is the conventional one in which a positive charge would move or tend to move.

This motion of electrons and holes, leaving the unbalanced ion charges which produce the electric field, occurs instantaneously on forming the junction. Immediately, the field reaches an equilibrium value which forbids further electron or hole migration across the p-n junction.

For convenience in explanation, the inevitable field and consequent p.d. (and remember that no external source of e.m.f. has yet been applied) is represented by a fictitious cell shown in dotted lines in Figure 3.4(b).

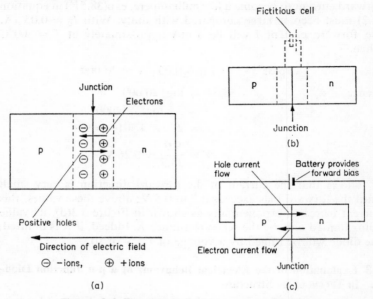

Figure 3.4. The electric field across a p-n junction

The region close to the p-n junction on both sides of it from which free current carriers have moved is called the *depletion region*.

When an external source of e.m.f. is used to set up a p.d. across the p-n junction which opposes the p.d. of the fictitious cell, current will flow and the junction is forward biased (Fig. 3.4(c)). This opposition corresponds to the positive terminal of the external voltage supply being connected to the p-type material and the negative terminal to the n-type.

Reversing the polarity of the applied potential difference provided across the p-n junction by the source of e.m.f. will clearly increase the electric field across the p-n junction and no majority carriers (electrons in n-type, holes in p-type) can flow through the p-n junction. However, the minority carriers (holes in n-type and

Semiconductor Diodes and Rectification

electrons in p-type) can flow. They constitute the reverse or leakage current. This leakage is much higher in germanium than in silicon.

The width of the depletion region (Figure 3.5) is increased on application of the reverse voltage. This is because the free current carriers are repelled from the p-n junction.

The n region and the p region separated by the depletion region form a parallel plate capacitor. The dielectric of this capacitor is the depletion region. The capacitance is increased if the area of cross-section of the p-n junction is increased and decreased if the depletion width is increased.

The fact that the capacitance of a junction diode is reduced by

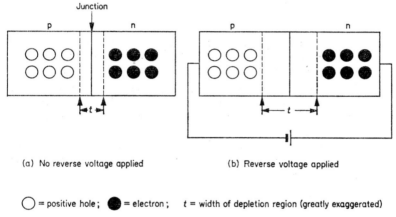

(a) No reverse voltage applied   (b) Reverse voltage applied

○ = positive hole;   ● = electron;   t = width of depletion region (greatly exaggerated)

Figure 3.5. The depletion region

the application of a reverse voltage (depletion region width increased) has been exploited in the development of the *varactor diode*.

If a p-n junction diode is needed for operation at very high frequencies, the capacitance must be very small, otherwise the capacitative reactance X_c ($X_c = 1/(2\pi f C)$, where C is the capacitance and f is the frequency) will be small and significant unwanted high frequency alternating currents will flow. For detectors (essentially rectifiers) of very high frequency alternating potentials of small amplitude, point contact diodes of very small p-n junction cross-section area are used.

3.4 Brief Notes on Relevant Aspects of Alternating Current

To examine the performance of various rectifying arrangements it is necessary to recall the following facts:

(a) An alternating current of single frequency f and pulsatance $\omega = 2\pi f$, is represented by

$$i = I_p \sin \omega t \tag{3.3}$$

where i is the instantaneous current at time t and I_p is the peak current.

(b) The root-mean-square (r.m.s.) current I of an alternating current represented by equation (3.3) is given by

$$I = I_p/\sqrt{2}$$

I may be defined as the magnitude of the direct current which produces in a given time the same heating effect in a resistance R as the alternating current of peak value I_p.
Similarly, for voltages,

$$V = V_p/\sqrt{2}$$

(c) Let V_p be the peak value obtained in a half-wave rectified voltage. The average value of this voltage is V_p/π. For a full-wave rectified voltage, the average value V_a is $2V_p/\pi$.

(d) For a diode, the *peak inverse voltage* (PIV) is the maximum reverse voltage that it can withstand during the non-conducting interval. Exceeding the PIV will cause the diode to fail prematurely.

Note from (b) and (c) that for a full-wave rectified voltage of peak value V_p, the average value V_a is related to the r.m.s. value V by

$$V_a = 2V_p/\pi = 0.64V_p = 0.64V\sqrt{2} = 0.9V$$

3.5 Half-wave Rectification

In the circuit used (Figure 3.6(a)) the primary winding of the mains transformer is connected across the a.c. mains (240 V r.m.s., 50 Hz). A fuse in series with this primary is a useful precaution. Across the secondary winding of this transformer is connected the load resistance R_L with the diode in series. Across R_L is produced a voltage which is unidirectional and consists of a series of pulses as shown in Figure 3.1(b).

With the diode connected as shown (the anode being joined to end A of the secondary winding and the cathode to end C of the load resistance R_L) it will conduct during each half-cycle when A is positive with respect to B, so there will be current through R_L. During the intervening half-cycles when A is negative with respect to B, the diode does not conduct (or rather conducts very poorly) so there is no or negligible current through R_L.

With a 50 Hz supply there will consequently be current pulses (each of half-cycle wave form) through R_L during, say, the first, third, fifth, seventh and so on hundredths of a second and corresponding voltage pulses developed across R_L. During the intervening second, fourth, sixth and so on hundredths of a second, the current through R_L will be zero for practical purposes and correspondingly no voltage will appear across R_L.

To demonstrate graphically the shape of the voltage pulses across R_L a cathode ray oscillograph (CRO) is connected across R_L. The input to the Y-plates of the oscillograph is best made via a

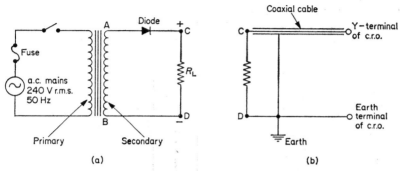

Figure 3.6. (a) Circuit diagram of a half-wave rectifier and (b) connection to a cathode ray oscillograph to demonstrate the waveform

coaxial cable connection. The central wire of this coaxial cable connects end C of the load resistance R_L to the Y terminal of the CRO. The outer screen or metal sheath of the coaxial cable is connected to end D of R_L and to the earth terminal of the CRO (Figure 3.6(b)).

Note that any point of the secondary circuit can be earthed as required because this secondary is isolated from the mains by the transformer.

Example 3.5

A half-wave rectifier circuit of the type shown in Figure 3.6(a) operates on an a.c. mains supply of 240 V r.m.s. with a step-down transformer of which the turns ratio T is $\frac{1}{3}$. Calculate (i) the peak inverse voltage across the diode and (ii) the average value of the output voltage across the load resistance.

The r.m.s. voltage across the secondary winding is
$$240 \times T = 240/3 = 80.$$

The peak secondary voltage $= 80\sqrt{2} = 113$. This will be the peak inverse voltage impressed across the diode during the non-conducting half-cycles. This is because the current through the load resistance is zero at these times and there is consequently no potential drop across this resistance. Hence the diode, when non-conducting, has to withstand the whole peak secondary voltage of 113 V.

The average output p.d. across the load resistance is given by

(peak secondary voltage)$/\pi = 113/\pi = 36$ V

It is here assumed that the resistance of the conducting diode is very small compared with the load resistance so that, when it is conducting, the potential drop across the diode is negligible.

3.6 Full-wave Rectification

The circuit (Figure 3.7) makes use of a mains transformer with a centre-tapped secondary winding, i.e. a terminal connection to the half-way point C in the winding is available so that the number of

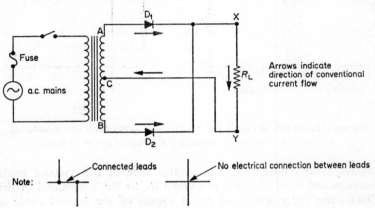

Figure 3.7. A full-wave rectifier unit

turns in the half AC equals that in half CB. End A of this secondary is connected via diode D_1 to one end X of the load resistance R_L. The other end B of the secondary is connected via a second similar diode D_2 *to the same point* X. The other end Y of the load resistance R_L is joined to the centre-tap C. Frequently, this point C is joined to earth.

The use of a centre-tapped secondary winding of a transformer is a simple and much used general method of obtaining two alternating voltages in anti-phase (exactly 180° out of phase). The potential at the centre-tap C is the reference point: this may or may not be earthed. Assume it is earthed for convenience so we may regard the potential of C as zero. When the alternating p.d. occurs across the whole secondary winding AB, at a particular instant of time suppose A is positive with respect to B. A will then be positive with respect to C, but B will be equally negative with respect to C. For example, if the instantaneous p.d. across AB is 100 V, that of A with respect to C will be +50 V, whereas that of B relative to C will be −50 V. This equal and opposite division will be maintained throughout: the potential of B with respect to C thus varies in anti-phase with that of A relative to C.

In the circuit of Figure 3.7, during a half-cycle when A is positive with respect to C, the diode D_1 conducts but as meanwhile B is negative with respect to C, diode D_2 does not conduct. In the immediately succeeding half-cycle, A will become negative relative to C whereas B becomes positive relative to C. Now the diode D_1 does not conduct whereas diode D_2 does. During both half-cycles note that the direction of the current through R_L is the same.

The unidirectional voltage across R_L thus varies with time as shown in Figure 3.1(c).

The average voltage V_a is $2V_p/\pi$, where V_p is the peak voltage across each half of the secondary winding.

The peak voltage across the whole secondary winding is $2V_p$. This will clearly be the peak inverse voltage across each diode.

If both diodes D_1 and D_2 were reversed, i.e. cathode of D_1 joined to A instead of anode and the same for D_2 in relation to B, the full-wave rectifier would still function. Now the centre tap C would be the positive terminal and current would flow in the opposite direction through the load resistance.

3.7 The Full-wave Rectifier Bridge Circuit

An alternative circuit to that of Figure 3.7 is one which dispenses with the centre-tap to the secondary winding and, instead, utilizes four diodes D_1, D_2, D_3 and D_4 in a bridge arrangement (Figure 3.8(a)).

In remembering how to draw this circuit note that the diodes in the four arms of the 'bridge' have their arrow heads all pointing 'upwards' (or they could all be 'downwards').

During those half-cycles of the sinusoidal voltage variation across the transformer secondary winding when A is positive with respect to B, diodes D_1 and D_2 conduct (but D_3 and D_4 do not). The appropriate part of the circuit clarifying this statement is drawn in Figure 3.8(b). During the intervening half-cycles when A is negative with respect to B, diodes D_3 and D_4 conduct (but D_1 and D_2 do not).

The voltage waveform which appears across the load resistance R_L (Figure 3.8(c)) has each half-cycle labelled with the appropriate two diodes which are forward biased. In Figure 3.8(a), the full arrows indicate the direction of the current flow during those half-cycles when A is positive with respect to B and the dotted arrows are for the intervening half-cycles when the polarity of A relative to B is reversed.

As with any full-wave rectifier, the average output voltage across R_L is $V_a = 2V_p/\pi$. As V_p in Figure 3.8(a) is the peak voltage across the full secondary winding and this is applied across two diodes in

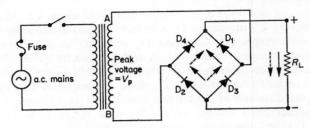

Full arrows indicate current flow when A+ w.r.t. B; dotted arrows, the current flow when A− w.r.t. B.

(a)

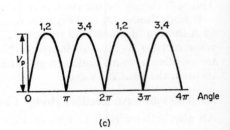

(b)   (c)

Figure 3.8. A full-wave rectifier bridge circuit

parallel, when these diodes are non-conducting, the peak inverse voltage across each is V_p.

Certain features of bridge rectifiers are worth noting:

(i) Four diodes are required instead of the two of a centre-tapped transformer secondary circuit, but the average output voltage is $2V_p/\pi$ where V_p is the peak voltage across the whole secondary winding and not only half of it.

(ii) With two diodes in series with R_L during conduction, the potential drop across them is twice that for one only. As the potential drop across a semiconductor rectifier when forward biased is negligible, the use of two rectifiers in series is generally of no consequence in this connection.

(iii) If any one diode is faulty so that it conducts when reverse biased, a second diode is always damaged because of the excessive current flow generated. It is good practice to check the performance of the individual diodes before the bridge rectifier is constructed.

(iv) The bridge rectifier is compact and often used in rectifier-type moving-coil instruments for a.c. measurements.

(v) Most manufacturers of semiconductor components market bridge rectifiers as compact units. The four diodes are mounted and inter-connected appropriately on the necessary heat sinks and tested. The bridge rectifier is therefore quickly connected as a unit component with two input terminals and two output terminals.

3.8 A Voltage Doubling Circuit

On occasions when a high d.c. voltage is required and the current demands are moderate, a voltage doubler circuit may be used. Whereas with a full-wave bridge rectifier circuit the average output voltage is $V_a = 2V_p/\pi$, where V_p is the peak secondary voltage, a voltage doubler circuit may be constructed which provides a quite steady, output voltage of $2V_p$, which is π times as great even though the same transformer is used.

In the voltage doubling circuit (Figure 3.9(a)) two equal diodes D_1 and D_2 are connected in the same direction in series. Across

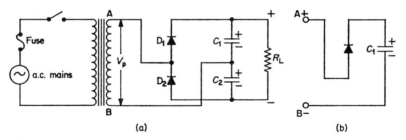

Figure 3.9. A voltage doubling circuit

them are connected two equal capacitances C_1 and C_2 in series and also the load resistance R_L. The transformer secondary winding is connected across the junction between D_1 and D_2 and the junction between C_1 and C_2.

If the load resistance R_L is large, corresponding to a small current drain on the circuit, the capacitances C_1 and C_2 in series will discharge through R_L at a much smaller rate than they are charged by the rectifiers D_1 and D_2.

During those half-cycles of the alternating potential difference across the transformer secondary AB when A is positive with respect to B, current flows through diode D_1 (but not through diode D_2) to charge up capacitance C_1. This is illustrated by part of the circuit drawn in Figure 3.9(b) in which the effect of the resistance R_L is omitted on the basis that the discharge current through it is comparatively very small.

A series of unidirectional pulses of current fed into a capacitor of capacitance C will, in time, provide a charge $Q = CV_p$, where V_p is the peak voltage obtaining at each pulse, presuming that the capacitor is not being discharged. This is rather like saying that a series of drops of water must eventually fill up a container if there is no hole in the container through which water can leak. If the current pulses are large enough and occur in rapid sequence additively, the capacitor voltage of V_p is quickly obtained.

During the intervening half-cycle when A is negative with respect to B, current flows through diode D_2 (but not through D_1), to charge up capacitance C_2.

Each of the capacitances C_1 and C_2 is therefore charged up to the peak voltage V_p across the transformer secondary. The polarity of the voltage across C_1 is the same as that across C_2. The p.d. set up across C_1 and C_2 in series, which equals that across the large resistance R_L, is therefore $2V_p$.

In practice there must be some discharge current through R_L. If this current is very small, the p.d. across R_L is $2V_p$ and very nearly constant. As R_L is made smaller, so the current through it increases, the voltage across it will drop and then rise again, so there will be fluctuations of this voltage. The extent of these fluctuations is clearly a question of balancing the rate of charge of C_1 and C_2 from the rectifiers against the rate of discharge through R_L.

3.9 An Alternative Voltage Doubling Circuit

In an alternative circuit to that of Figure 3.9, a capacitance C_1 is connected in series with the transformer secondary and a diode D_1, then across D_1 is connected a second diode D_2 in series with a second capacitor C_2. The load resistance R_L is across the capacitance C_2 (Figure 3.10).

During those half-cycles when B is positive with respect to A, diode D_1 conducts (but D_2 does not) so that capacitor C_1 is charged to the peak transformer secondary voltage V_p. Note that plate X of capacitance C_1 becomes positive with respect to A.

During the intervening half-cycles when A is positive with respect to B, diode D_2 conducts (but D_1 does not). The p.d. across D_2 and C_2 in series is now that across AB plus that across C_1: it is therefore $2V_p$. Hence C_2 becomes charged to a p.d. of $2V_p$.

As in the circuit of Figure 3.9(a), this action requires that the current drain (rate of discharge) through R_L is small compared with the rate of charging of C_1 and C_2 so that these capacitors can act as reservoirs of charge.

It is readily appreciated that in both these voltage doubler circuits (Figures 3.9 and 3.10) the peak inverse voltage across each diode is $2V_p$.

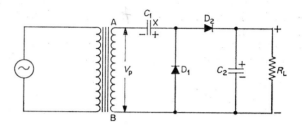

Figure 3.10. Voltage doubler alternative to that of Figure 3.9

3.10 A Voltage Quadrupler Circuit

Two circuits of the voltage doubler type shown in Figure 3.10 may be arranged as in Figure 3.11 to form a quadrupler circuit. Provided that the load current is very small (R_L very large) a p.d. across the load of $4V_p$ can be obtained.

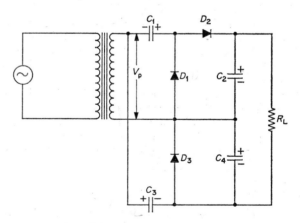

Figure 3.11. A voltage quadrupler circuit

3.11 Smoothing Circuits

After rectification (Sections 3.5 and 3.6) the output voltage, although unidirectional, consists of pulses. Although these voltage pulses are approximately half of a sinusoidal wave form, the fluctuations which occur about the average voltage V_a are not sinusoidal. Nevertheless, the fluctuations repeat themselves exactly at regular intervals of time. The mathematical method of Fourier analysis (beyond the scope of this text) enables periodic waveforms of most shapes to be expressed in terms of sine waves of a fundamental frequency (first harmonic) and higher harmonics (second, third, fourth etc. of 2 times, 3 times, 4 times etc. respectively the fundamental frequency).

The rectified sine wave can thereby be shown to be represented by the summation of a number of simple components which include:

(i) A steady voltage (or current, if current is the concern) of value equal to V_a, the average value. This average is V_p/π for half-wave and $2V_p/\pi$ for full-wave rectification (section 3.4).

(ii) An alternating voltage of which the frequency f is the same as that of the original alternating supply (at 50 Hz for the a.c. mains) which is rectified. This frequency f is the fundamental or the first harmonic.

(iii) Alternating voltages of frequencies $2f$, $3f$, $4f$ etc. which are correspondingly the second, third, fourth harmonics etc.

In an introductory account of smoothing circuits it is adequate to ignore the second and higher harmonic components of frequencies nf where n is an integer of two or more. Indeed, in practice, if the smoothing method is able to deal with the fundamental frequency component, it is generally more readily able to cope with the higher harmonic components.

What is required of a smoothing circuit, therefore, is a means of reducing significantly the alternating voltage of frequency f in the rectified output. This is achieved by a filter of which the simplest ones depend on a capacitor or an inductance. The filter has to be able to pass readily the d.c. component of the rectifier output and block the alternating component.

3.12 The Simple Capacitor Smoothing Filter

A rectifier circuit such as is described in sections 3.5, 3.6 and 3.7 delivers across a load resistance R_L either a half-wave or full-wave rectified output, depending on its design. Basing the discussion on the half-wave rectifier, the circuit of Figure 3.6(a) with the waveform shown in Figure 3.1(b) is typical.

The object of the smoothing filter is to render steady the pulsating voltage across R_L. To achieve this with a simple capacitor filter, a capacitance C is connected across R_L and a safety resistance R_s is

Semiconductor Diodes and Rectification

inserted in series between the diode and the capacitor to limit current surges.

The circuit is then as represented in Figure 3.12(a); the alternating input voltage has a peak value of V_p, that across the transformer secondary (the transformer is omitted in this diagram).

The capacitance C is charged by the rectified current pulses from the diode; it discharges through R_L. The rate at which it charges depends on the size of the current pulses passed through the diode, which is decided by the current capacity of this diode. (It is assumed that the transformer used is capable of providing current of the

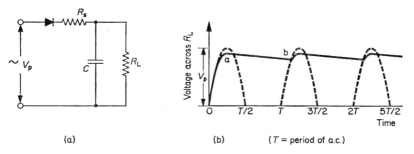

Figure 3.12. A simple capacitor filter used with a half-wave rectifier

magnitude needed.) The rate at which C discharges decreases as the time constant CR_L increases. It is clearly important that a steady charge is maintained within C, which acts as a reservoir capacitor. To avoid too rapid a discharge rate of C, either C or/and R_L must be big.

Initially, the capacitance C is discharged. When the circuit is first switched on, C may well charge up so rapidly that the current pulses through the diode are so large as to destroy it. This is avoided by the use of the safety resistor R_s, which limits the magnitude of charging current pulses.

Suppose that CR_L is large. This capacitance C will then charge up almost to the peak voltage V_p during the first conducting half-cycle (Figure 3.12(b)).

During the immediately succeeding half-cycle, the charging current is zero (the diode is reverse biased). The voltage V_p across C will therefore decrease exponentially because of the discharge through R_L. If CR_L is large enough, the decrease of voltage across C and R_L in parallel will be a small fraction of V_p during one period. Consequently the output voltage across C and R_L falls by only a small amount along ab and then, at b, it is restored almost to V_p again by the next charging half-cycle.

The voltage fluctuations (known as the *ripple voltage*) across C and R_L are therefore reduced to a small fraction of the magnitude of V_p that prevailed without the capacitance C. The percentage ripple is smaller the larger is the time constant CR_L.

For a full-wave rectifier, the discharge time is only half that for the half-wave so, other factors being the same, the percentage ripple is halved (Figure 3.13), and the ripple frequency is doubled.

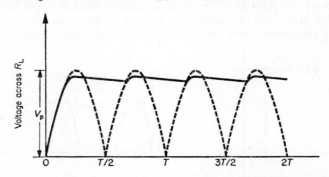

Figure 3.13. Full-wave rectified waveform showing the effect of the use of a smoothing capacitor

Another way of looking at the action of the smoothing capacitor is to consider that R_s ($\ll R_L$) and C are in series across the rectified a.c. supply which is equivalent to a steady voltage and an alternating voltage of frequency f (section 3.11). The reactance $X_C = 1/(2\pi fC)$ of the capacitance C is much less than the resistance R_L if C is large enough. The steady current output from the rectifier cannot pass through C so establishes a steady voltage across R_L. The alternating current component takes the much easier path through C so the fraction of this a.c. through R_L is much the smaller. Consequently, the voltage across R_L is the steady component subjected to a small alternating voltage: the ripple voltage

Example 3.12(a)

A capacitor of capacitance 100 μF *is used to smooth the output from a half-wave rectifier. The transformer secondary output is* 40 *V r.m.s.,* 50 *Hz, and the average rectified current (load current) passed by the diode is* 15 *mA. Calculate the average output voltage and the peak value of the ripple voltage. Draw a graph showing the way in which the output voltage varies with time.*

The peak voltage across the secondary winding is $\sqrt{2} \times 40 = 56$ V. The capacitor will charge to a maximum of very nearly 56 V during the conducting half-cycles and will discharge over a time of one cycle, which is 0.02 s.

The current through the load resistance = 0.015 A so that during 0.02 s, the quantity of electricity discharged is $0.02 \times 0.015 = 3 \times 10^{-4}$ C. The loss of charge of 3×10^{-4} C by the 100 µF capacitor must correspond to a fall of the p.d. across it given by

$$\frac{3 \times 10^{-4}}{100 \times 10^{-6}} = 3 \text{ V}.$$

so the minimum voltage across the capacitor is $56 - 3 = 53$ V.

The average voltage across the capacitor, or average output voltage, is therefore $(56 \times 53)/2 = 54.5$ V. The peak ripple voltage is half the fall of the capacitor voltage during one cycle and is hence **1.5 V**. These results are illustrated by the graph of Figure 3.14.

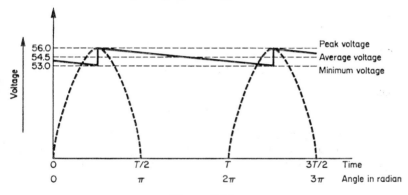

Figure 3.14.

Note: Before the circuit of example 3.12(a) is switched on the 100µF capacitor is uncharged When switched on, this capacitor is charged to nearly 56 V when the charge it holds will be

$$56 \times 100 \times 10^{-6} = 56 \times 10^{-4} \text{C}.$$

This charge could be established in the first millisecond if the transformer winding impedance were very low. The instantaneous current through the diode would then be

$$56 \times 10^{-4}/10^{-3} = 5.6 \text{ A}.$$

This is an excessive current for a diode only required to pass 15 mA in continuous operation. Though 1 ms is perhaps an underestimate of the time in some practice, this calculation nevertheless illustrates the importance of two factors One is that the instantaneous current rating of a diode is an important characteristic to be noted. The second is that a safety resistance R_s is often a necessity to prevent damage when the circuit is first switched on.

Example 3.12(b)

A full-wave rectifier operating from a 50 Hz supply provides a peak output of 33 V. A load resistance of 10 kΩ is put across this output with a reservoir capacitor in parallel. Calculate the capacitance of this capacitor required if the ripple voltage is to have a peak value of 1 V.

The average voltage across the capacitor is 32 V. The average load current is

$$32/10^4 = 32 \times 10^{-4}\,\text{A} = 3.2\,\text{mA}$$

The capacitor, on discharging for a time of half a cycle, which is 10^{-2} s, loses a charge Q of $3.2 \times 10^{-3} \times 10^{-2}$ C. The change of the p.d. across the capacitor during this time is to be 2 V for a peak ripple voltage of 1 V. Hence

$$2 = Q/C$$

where C is the capacitance required.

$$C = Q/2 = 3.2 \times 10^{-5}/2 = 16 \times 10^{-6}\,\text{F}$$

i.e.

$$C = 16\,\mu\text{F}.$$

3.13 Observations on Simple Capacitor Smoothing

This method is simple and offers the advantages of a high output voltage and satisfactory smoothing if the load current is small (i e R_L is large) The disadvantages are:

(i) the ripple voltage increases with the load current;
(ii) the output voltage drops significantly as the load current increases, so the voltage regulation is poor.

Nevertheless, most small power packs use silicon p-n junction diodes and simple capacitor filters. To improve the voltage stability, a Zener diode (section 3.15) or a series control transistor (section 4.15) is employed.

3.14 The Simple Choke Filter

A choke is an inductance L which, at the frequency f of the a.c. supply concerned, offers a reactance $2\pi fL$ which is much greater than its resistance R so that it impedes considerably (chokes) the flow of a.c. but not significantly that of d.c.

A simple choke filter is not often used for smoothing in small power packs but is of particular value when large output currents are required.

In a full-wave rectifier with a choke coil L connected in series with the load resistance R_L (Figure 3.15) the average d.c. voltage appearing across R_L is $2V_p/\pi$ where V_p is the peak voltage across the

Semiconductor Diodes and Rectification

transformer secondary and the ohmic resistance R of the choke is very low.

The choke of inductance L offers a high impedance to the flow of the alternating current component of the rectified output from the diodes. With a main supply of frequency 50 Hz to the transformer primary, the a.c. component involved has a frequency of 100 Hz.

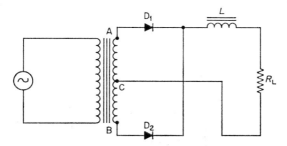

Figure 3.15. A simple choke filter used with a full-wave rectifier

The impedance of the choke (neglecting its ohmic resistance) is equal to its reactance $2\pi fL$. The fraction of the a.c. component from the rectifiers which appears across the load resistance R_L is therefore

$$R_L/\sqrt{[R_L^2 + (2\pi fL)^2]}$$

This expression shows that the smoothing is more effective the smaller the value of R_L and hence the greater the load current.

The choke or inductor L also provides a smoothing effect because when the current through it decreases, the magnetic flux around it decreases causing an induced e.m.f. in the correct direction (by Lenz's law) to offset this current decrease and *vice versa*, when the current through it increases, the induced e.m.f. brought about by the increasing magnetic flux opposes the current rise.

In comparing the choke and capacitor filters as smoothing devices it is noted that:

(i) Unlike the capacitor filter, the choke is of less value with a half-wave rectifier because it cannot so readily result in a continuous output.

(ii) With a full-wave rectifier, the average output voltage from the choke filter is less than from the capacitor filter.

(iii) The greater the load current, the more effective is the smoothing with a choke filter. The opposite is the case with a capacitor filter.

3.15 Zener or Voltage Regulator Diodes

Zener diodes (also known as voltage regulator diodes or as reference voltage diodes) are silicon p-n junctions created by the diffusion (section 2.6) or alloy method (section 2.5) which are operated at a reverse bias value which is just beyond the junction breakdown voltage. The circuit symbol is shown in Figure 3.16(a). To establish reverse bias the external d.c. supply is connected so that the p-type

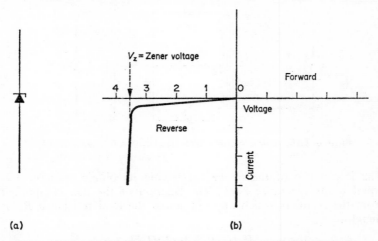

Figure 3.16. The Zener diode: (a) the circuit symbol; (b) the reverse bias characteristic

material (the 'anode') is negative and the n-type material (the 'cathode') is positive.

The reverse bias characteristic (Figure 3.16(b)) is like that of any conventional p-n junction diode, as shown in Figure 3.3(d). The Zener voltage V_Z is that at which the abrupt change in slope of the reverse bias characteristic occurs. When the applied reverse bias voltage V is less than V_Z, the resistance of the device is about 1 MΩ, and may be assumed to be infinite for most practical purposes. When V is increased to slightly beyond V_Z, the reverse current through the diode increases very rapidly. The current is now limited only by the external supply circuit. Hence in a circuit used to plot this characteristic and in the use of the Zener diode to provide a reference voltage, a series resistor must be included to prevent excessive current flow. Excess current would cause overheating and ultimate failure of the junction.

When a reverse bias is applied across a p-n junction the width of the depletion region increases because free current carriers are repelled from the junction (section 3.3). The width of the depletion region is inversely proportional to the level of doping, which is equivalent to saying that this width is proportional to the resistivity of the doped semiconductor material.

If low-resistivity silicon (that with a high level of doping) is used, the depletion region is very narrow. When a reverse bias is applied the electric field strength across the depletion region about the p-n junction is therefore very large. If this reverse bias is increased to the Zener voltage V_Z, the electric field is so high that electrons are torn away from their normal locations about the atomic nuclei in the material. These electrons contribute massively to the current so that Zener breakdown results.

Zener diodes are thus p-n diodes with special depletion region characteristics, usually such that the Zener breakdown occurs below 5 V. The breakdown has to occur under controlled reversible conditions. Thus, if V exceeds V_Z breakdown occurs; when V is then reduced below V_Z, the normal high impedance p-n characteristic is restored, and this can occur repeatedly, indefinitely and reliably.

This breakdown mechanism is temperature dependent. On the application of a strong electric field, atoms release electrons to become ions more readily as the temperature is increased. Therefore Zener diodes with V_Z up to 5 V have a negative temperature coefficient: V_Z decreases somewhat as the temperature is increased.

With p-n junction diodes made from higher-resistivity doped silicon, the depletion layer is wider. Correspondingly, a higher p.d. across the depletion layer is needed to produce a given electric field strength. Now the minority carriers which constitute the reverse bias current can acquire sufficient energy to ionize lattice atoms on collision because they can undergo an adequate potential drop even though the electric field strength is not high enough to produce a Zener breakdown. The electrons produced by such ionization are themselves accelerated in turn and cause further ionization. This cumulative phenomenon results in a current avalanche.

There are consequently two different mechanisms: Zener breakdown proper due to an intense electric field; and ionization by collision resulting in a current avalanche. The former predominates with narrow depletion regions between low resistivity material for which V_Z is below 5 V. The latter—the avalanche effect—predominates with wider depletion regions between high resistivity material for which V_Z is above 5 V.

Nevertheless, voltage regulator diodes of both types are made and both are usually called Zener diodes, despite the fact that those with a breakdown voltage greater than 5 V do not operate primarily as a result of true Zener breakdown. In the depletion region, the mobilities of the current carriers decreases with increases of temperature. The avalanche breakdown phenomenon hence exhibits a positive temperature coefficient.

In general, therefore, Zener diodes with $V_z < 5$ V have negative temperature coefficients and those with $V_z > 5$ V have positive temperature coefficients. Zener diodes at which Zener voltages are between 5 and 6 can be selected to have effectively a zero temperature coefficient.

3.16 Zener Diodes in Voltage Regulator Circuits

A voltage regulator is a device which, when connected across a voltage supply source to a load, maintains constant the potential difference across the load despite fluctuations in the value of the load resistance or of the output voltage from the supply source.

A device is able to fulfil this purpose if it is such that, when the voltage across it exceeds some critical value, this voltage is independent of the current through the device.

The reverse bias characteristic of the Zener diode (Figure 3.16(b)) satisfies excellently this requirement. Once the Zener voltage V_Z has been reached, the characteristic is steep and linear. The slope of this linear portion is called the *slope resistance* of the diode.

Slope resistance

$$= \frac{\text{a small increase in the reverse bias voltage}}{\text{the corresponding increase in the current}}$$

$$= \Delta V/\Delta I$$

The lower the slope resistance, the more closely does the diode approach the ideal with a slope resistance of zero.

In choosing a voltage regulator diode, the manufacturer's data will provide:

(i) I_Z, the reverse current at which the characteristic was measured. This is obviously chosen so that the operating point on the characteristic (Figure 3.16(b)) is clear of the turnover point or 'knee'.

(ii) V_Z, the Zener voltage, e.g. minimum 5.3 V; maximum 5.9 V. There is always a 5 to 10 per cent tolerance in the manufacture of Zener diodes: they are not all exactly alike, but any one diode will always have a constant specific value of V_z.

(iii) The typical temperature coefficient of the breakdown voltage expressed as per cent K^{-1}. In the voltage range from 5 to 6, these coefficients would be very small, i.e. <0.001 per cent K^{-1}.

(iv) The slope resistance, e.g. 55 Ω

(v) $I_{z\ max}$ the maximum Zener current allowable at 40°C, e.g. 50 mA.

(vi) P_{max}, the maximum allowable power dissipation at 40°C, e.g. 300 mW.

Knowing this data and its current-voltage characteristic, it is necessary to decide the minimum current $I_{z\ min}$ to be clear of the 'knee'. Thus it might be decided that $I_{z\ min} = 1$ mA and that the Zener voltage $V_z = 5.8$ V.

3.17 A Simple Zener Voltage Stabilizer

In the circuit of Figure 3.17, V_1 is the voltage from a power pack (e.g. a rectifier with capacitor smoothing), which may vary, and R_L is the load resistance (which may also vary) across which it is required to supply a constant voltage, irrespective of changes of V_1 or of the load current I_L through R_L.

This is achieved by the use of a resistance R_1 in series with a Zener diode across the voltage supply V_1 and the connection of R_L

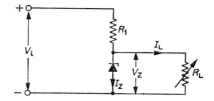

Figure 3.17. A simple Zener stabilizer circuit

across the Zener diode. The value of R_1 is selected so that the voltage across the Zener diode (and so across R_L) is the Zener voltage V_z.

The voltage V_z across the Zener diode cannot alter but only the current I_z through it. An increase of the load current I_L, which would normally tend to reduce V_z because of the increased voltage drop across R_1, causes reduced current I_z to flow through the Zener diode so that the total current $(I_L + I_z)$ through R_1 remains unchanged.

When I_L is a maximum, a minimum allowable current $I_{z\ min}$ must still flow through the Zener diode, and when I_L is zero the maximum current that flows through the Zener diode must not damage it.

The design considerations involved are illustrated by an example. Suppose a Zener diode having the typical characteristics given in section 3.16 is used. It is required to design a stabilizer which provides 5.8 V. It is assumed that the load current can vary between 0 and 10 mA and that the supply voltage V_1 is nominally 9 V but never exceeds 10 V and must always be greater than 1 V above V_Z, i.e > 6.8 V.

When the load current I_L falls to zero, the maximum current possible that flows through the Zener diode is $(I_{L\,max} + I_{Z\,min})$ where $I_{L\,max}$ is the maximum load current encountered. This is 10 mA plus 1 mA (section 3.16), giving 11 mA.

Suppose the minimum p.d. across the series resistance R_1 is chosen to be 1.5 V. Then

$$R_1 = \frac{1.5}{11 \times 10^{-3}} = 136\ \Omega$$

The nearest preferred value of R_1 selected from amongst the resistor values available from manufacturers is 150 Ω. Using this value, the maximum current $I_{Z\,max}$ which flows through the Zener diode is when the input voltage V_1 is at its maximum of 10 V and the load current I_L is zero. As $V_Z = 5.8$ V, so

$$I_{Z\,max} = \frac{10 - 5.8}{150} = 0.028\ \text{A} = 28\ \text{mA}$$

This gives the maximum power dissipation in the Zener diode,

$$V_Z I_Z = 5.8 \times 0.028\ \text{W} = 162\ \text{mW}$$

which is well within the rated value of 300 mW.

3.18 Zener Diodes in Series and in Parallel

The use of three Zener diodes in series (Figure 3.18) enables six different stable output voltages to be obtained. They are V_{Z1},

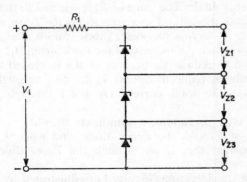

Figure 3.18. Three Zener diodes in series

V_{Z2}, V_{Z3}, ($V_{Z1} + V_{Z2}$), ($V_{Z2} + V_{Z3}$) and ($V_{Z1} + V_{Z2} + V_{Z3}$). The current through any one of these series connected Zener diodes must not be allowed to fall below $I_{Z\ min}$.

The connection of Zener diodes in parallel (Figure 3.19) enables

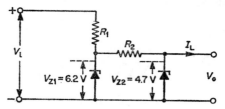

Figure 3.19. Two Zener diodes in parallel to give much improved stability

very good stability to be achieved and is especially recommended if the supply voltage V_i contains an alternating ripple component. The design is as in section 3.17 with each Zener diode being treated as a single stage. In a circuit such as that of Figure 3.19, it must, of course, be ensured that the Zener voltage for diode 1 exceeds that of diode 2, i.e. $V_{Z1} > V_{Z2}$ and where V_{Z2} equals V_0.

3.19 A Variable Stabilized Supply

By making resistor R_2 of Figure 3.19 a variable component, two Zener diodes in parallel may be used in a circuit which provides a

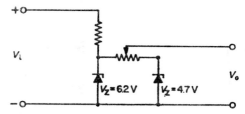

Figure 3.20. Two Zener diodes in parallel in a circuit which provides an adjustable stabilized voltage

variable output which is stabilized between 4.7 V and 6.2 V, for example (Figure 3.20). The stability is not so good as that obtained with the circuit of Figure 3.19.

3.20 Further Applications of Zener Diodes

Two more useful applications of Zener diodes are:
    (i) To protect a voltmeter from excessive voltage: a Zener diode with V_Z at approximately twice that for full-scale deflection is conveniently used (Figure 3.21(a)).

(ii) As a means of limiting the height of a voltage pulse (Figure 3.21(b)). For example, the tops may be clipped from a unidirectional sinusoidal wave form voltage supply to produce a series of pulses of roughly rectangular wave form of constant peak voltage values.

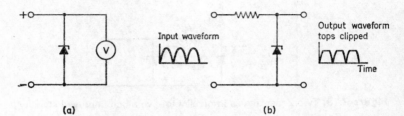

Figure 3.21. (a) Use of a Zener diode to protect a voltmeter from overload and (b) a Zener diode limiter

3.21 Shunting a Current Meter to Provide a Non-Linear Scale

An interesting application of the silicon diode (*not* the Zener diode) is to use it in series with a resistance to shunt a microammeter. It is common practice to connect a low resistance in parallel with, say, a $0 \rightarrow 100$ μA meter to enable the meter to be used in a higher

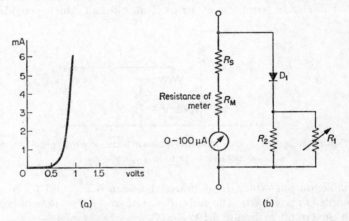

Figure 3.22. Arranging a current meter to have a non-linear scale

current range, say $0 \rightarrow 10$ mA. If this is done with the usual moving-coil meter, both the scales (with and without the shunt) are linear.

A silicon diode, of which the voltage–current characteristic is shown in Figure 3.22(a), in the shunt circuit (Figure 3.22(b)) enables the scale to be highly compressed over a certain current range whilst remaining approximately linear at each end, say between 0 and 15 μA at the lower end and between 1 and 10 mA at the high end. This provides a shunted meter which is convenient in many applications in that, on the one scale, a current can be recorded with an accuracy of about 5 per cent up to 15 μA and from 1 to 10 mA, and also give an indication of the current between these extremes.

Initially a resistance R_s is connected in series with the resistance R_M of the meter to make $(R_M + R_s)$ equal to 12 kΩ approximately. In a suitable circuit with a current of 10 mA flowing, the variable resistance R_1 is adjusted to give a full-scale deflection on the 100 μA meter. With R_1 fixed at this value, the scale of the meter is calibrated against that of a good standard meter (e.g. an Avometer model 8). In general, the resistance R_M of the microammeter is approximately 1000 Ω so that the value of R_s required in approximately 11 kΩ. Any low-power silicon diode will serve for D1 in Figure 3.22(b), whilst R_2 could be 82 Ω and R_1 a variable resistance of maximum value 1 kΩ.

The behaviour of the meter incorporating the non-linear shunt can be understood in a qualitative fashion from the diode characteristic (Figure 3.22(a)). When the voltage drop across the diode in the forward biased direction is small, say 0.1 V, the resistance of the diode is very high so negligible shunting of the meter occurs. The fact that the meter resistance including R_s is 12 kΩ in this small current range is of no significance because the total resistance in any circuit operating in this current range (say, 0 → 15 μA) is usually very large in comparison. As the voltage across the resistance of 12 kΩ $(R_s + R_M)$ increases, the diode current increases rapidly so the scale is very compressed. As the voltage across the diode increases, its effective resistance becomes very small and the 100 μA meter is shunted in the normal way.

Such a modified meter is useful in the simple transistor tester section 4.17). However, when reversing the polarity of the battery to check an n-p-n transistor, the meter terminals must also be reversed.

Again, when using a centre-zero meter to detect the balance condition in a Wheatstone bridge, it is usual to protect the meter movement by including a large series resistance which is short-circuited by a switch as the balance point is approached. Two silicon diodes can be used (to modify the meter movement in each direction) so that even an out-of-balance current of 10 mA would not damage the movement of a microammeter. In addition an out-of-balance current of 1 μA could still be detected, though with slightly less sensitivity than that produced by the microammeter alone.

Example 3.21

Referring to the circuit of Figure 3.22(b), calculate the effective resistance of the diode when a full-scale deflection of the 100 μA meter corresponds to a current of 10 mA, when $R_2 = 82\ \Omega$, $R_1 = 656\ \Omega$ and $R_s + R_M = 12\ k\Omega$.

The appropriate circuit (Figure 3.23) includes the resistance R_D of the diode. The potential difference across $(R_s + R_M)$ is that due to 100 μA through 12 kΩ, i.e. it is $(12 \times 10^3 \times 10^{-4})V = 1.2\ V$.

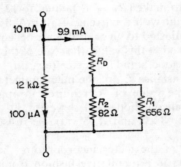

Figure 3.23.

The current through the shunt must be 10 mA − 100 μA = 9.9 mA. As the p.d. across the shunt must also be 1.2 V, it follows that the total resistance of the shunt must be

$$\frac{1.2}{9.9 \times 10^{-3}} = 120\ \Omega$$

This must equal R_D is series with the combination of R_2 and R_1 in parallel. The resistance of this parallel combination is R given by

$$\frac{1}{R} = \frac{1}{82} + \frac{1}{656} = \frac{8 \times 1}{656} = \frac{9}{656}$$

$$R = 656/9 = 72\ \Omega$$

As $R_D + R = 120\ \Omega$, hence R_D is 48 Ω.

Exercise 3

1. Explain the rectifying action of a p-n junction and compare the characteristic of such a junction with that of a vacuum diode.
(A.E.B., part).

2. A forward-biased p-n junction diode at 300 K has a characteristic of which the voltage and current values are given in the table below:

Semiconductor Diodes and Rectification

Forward voltage (volt)	0.05	0.10	0.15	0.20	0.25	0.30
Forward current (ampere)	10^{-6}	5×10^{-6}	2×10^{-5}	10^{-4}	4×10^{-4}	1.5×10^{-3}

The relationship between the diode current I and the voltage V across the diode is given by the equation

$$I = I_0[\exp(eV/kT) - 1]$$

where e is the electronic charge, k is the Boltzmann constant, T is the absolute temperature and I_0 is a constant.

Present the values given in the table in the form of a graph suitable to verify this equation and calculate values for I_0 and for k. (Electronic charge $e = 1.6 \times 10^{-19}$C).

3. It is required to identify the electrodes of an unmarked p-n junction diode and also to decide if it is made from germanium or silicon. Describe any experiment or tests necessary to achieve this.
(A.E.B., part).

4. Draw circuit diagrams and explain the function of the components in a full-wave rectifier circuit using
   (a) a centre-tapped transformer;
   (b) a bridge-type rectifier.
   Outline the advantages and disadvantages of each.

5. In connection with a full-wave rectified wave form, define and establish relations between the following values:
   (a) peak voltage;
   (b) root-mean-square voltage;
   (c) average voltage.
   Draw the circuit diagram of a voltage doubling circuit and explain its operation.

6. A simple half-wave rectifier is connected to a sinusoidal a.c. supply of 100 V r.m.s. at 50 Hz, and is loaded with a resistance of 1000 Ω. Draw a diagram of the voltage waveform across the load resistance and calculate the peak value of the rectified current, assuming that the diode used is ideal.

7. A half-wave rectifier with a smoothing filter in the form of a capacitor of 8 μF is required to supply an average load current of 20 mA at an average voltage of 250. This mains supply is 240 V r.m.s. at 50 Hz. Calculate the transformer ratio required.

8. Compare the behaviour of a simple capacitor filter with that of a simple choke filter in smoothing a rectified supply.

9. Explain the rectifying action of a silicon p-n junction diode. Draw circuit diagrams of full-wave rectifiers using (i) a centre-tapped

transformer and (ii) a bridge type circuit. Explain the action of a simple reservoir smoothing-capacitor.

A full-wave rectifier of peak output voltage 25 V contains a simple reservoir smoothing capacitor of capacitance C. If the load on the rectifier is 4 kΩ calculate the value of C required to ensure that the peak ripple voltage does not exceed 1.0 V. Frequency = 50Hz. (A.E.B.)

10. Explain the terms *depletion layer*, *minority carriers* and *reverse saturation current*. What factors determine the magnitude of the Zener voltage and the temperature coefficient of a reference voltage diode?

11. Distinguish between Zener and avalanche breakdown in voltage regulator diodes.

Explain, with the aid of a circuit diagram, how a Zener diode can be used to provide a steady voltage across a load, despite the fact that both the input voltage and the load resistance can vary.

12. It is required to make the response non-linear of a microammeter which has a full-scale deflection of 100 µA. The object is to achieve a scale which is linear in the region 0–15 µA and also that the meter will indicate currents of up to 10 mA without the use of a range switch. Explain how this can be achieved.

4 The junction transistor

4.1 Structure and Behaviour

The junction transistor—a very important control device in electronics—consists of two p-n junctions positioned close together within a single crystal slice. The region common to the two junctions, called the *base*, may be either of n-type semiconductor material or of p-type. In the first case, the n-type base region is flanked on each side with p-type material, forming a p-n-p junction transistor. In the second case, the p-type base region is flanked by n-type material, forming an n-p-n junction transistor. The thin central region—the base (b)—is of high resistivity material sandwiched between more heavily doped material comprising the *emitter* (e) on one side and the *collector* (c) on the other side. At present it is easier to mass-produce p-n-p transistors in germanium and n-p-n ones in silicon. These are the most commonly encountered junction transistors.

The structures are illustrated in outline in Figure 4.1(a) The distinction between the conventional circuit symbols of the two types (Figure 4.1(b)) is made by the direction of the arrowhead joining the emitter to the base. This arrowhead points in the direction of *conventional* current flow (i.e. from positive to negative, opposite to that of electrons). Thus, in the p-n-p transistor, the arrowhead is directed from emitter to base because positive charge flows readily from p to n across the junction, whereas in the n-p-n transistor, the arrowhead points from base to emitter.

The two junctions formed are the emitter-base junction (J_{eb}) and the collector-base junction (J_{cb}). Each of these junctions behaves in a similar way to the p-n junctions already described in Chapter 3. However, provided that the junctions are correctly biased (i.e. have small voltages of the correct polarity across them), additional and most useful behaviour results from these junctions being so close together, separated only by the thin base region.

To obtain transistor action the emitter-base region must be forward biased and the collector-base junction reverse biased.

To obtain forward bias across the emitter-base region of a p-n-p transistor, the base must be made negative with respect to the emitter; positive holes then flow readily from the p-type emitter to the n-type base across the junction (J_{eb}). For an n-p-n type transistor, the base has to be positive with respect to the emitter to provide forward bias so that now electrons flow readily from emitter to base.

The behaviour of the p-n-p transistor is considered in detail. That of the n-p-n type will be apparent, bearing in mind that the applied voltages will be of opposite polarity and the majority carriers will be of opposite charge.

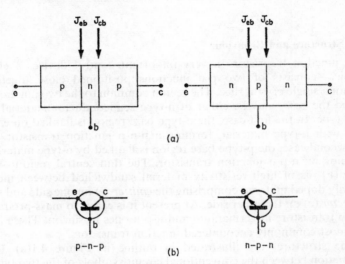

Figure 4.1. (a) Schematic diagrams of p-n-p and n-p-n junction transistors and (b) circuit symbols for these transistors

On applying forward bias to the emitter-base junction of a p-n-p transistor, the consequent movement of positive holes from the emitter into the n-type base region causes more holes to move into or be injected into the emitter to maintain equilibrium. Once in the base region, the holes do not have far to travel by diffusion through the thin base before they are accelerated across the collector-base junction (J_{cb}) to the collector electrode; this is due to the reverse bias across J_{cb} which, with the collector, is negative with respect to the base.

Some holes will combine with electrons in the base region and so are unable to reach the collector. Electrons which move into the base to compensate for this loss due to recombination will constitute the base current I_b.

The greater the reverse bias across the collector-base junction, the further will the depletion layer (section 3.3) extend into the n-type base. This effectively decreases the width of the base and reduces the number of holes lost by recombination.

To provide current flow in a forward biased p-n junction in germanium a p.d. of about 0.2 V is required; for silicon about 0.5 V is needed (section 3.2).

A junction transistor may be connected in one of three ways: *common-base*, *common-emitter* or *common-collector*. As the last of these is the least used it is not considered further here. Common-emitter connection (Figure 4.2(a)) is the most widely used method;

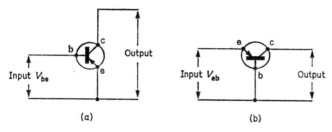

Figure 4.2. (a) Common-emitter connection and (b) common-base connection

the common-base connection (Figure 4.2(b)) is not so versatile. The bias across base-emitter is V_{be}, across collector-emitter it is V_{ce} and across collector-base, C_{cb}. The electrode stated to be common is in fact that one which is common to both the input and the output circuits.

4.2 The Characteristics of a p-n-p Transistor in Common-Emitter Connection

The circuit of Figure 4.3(a) enables the characteristics of p-n-p transistors in common emitter connection to be plotted. An example is an ACY21 (a p-n-p germanium transistor manufactured by Mullard Ltd.). The base connections are shown in Figure 4.3(b). *Voltage supplies should not be connected until the polarity has been checked against the circuit diagram.* The transistor could be permanently damaged if a junction bias were incorrect.

The voltmeters used to record V_{be} and V_{ce} should have a high resistance of about 10 MΩ*. If multi-range instruments are used for the current measurements (I_b and I_c), their range should not be altered during an experiment.

The meter used to record the base current I_b may have a full

---

* The simple FET voltmeter described in 'Educational Projects in Electronics' (Mullard Ltd.) is ideal; see section 5.12.

scale deflection (f.s.d.) of 250 μA, whilst that for the collector current I_c may have an f.s.d. of 25 mA.

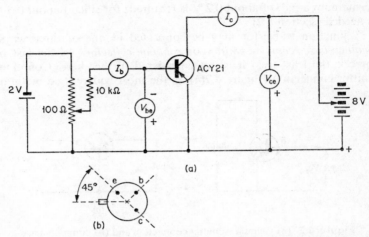

Figure 4.3. (a) Circuit for plotting the characteristics of a p-n-p transistor in common-emitter connection and (b) base contacts of the ACY21, viewed from the underside; the base electrode (b) is connected internally to the metal case

The *input characteristics* (Figure 4.4) are those of the base-emitter voltage (input voltage) V_{be} plotted against the base current (input current) I_b for a number of values of the collector-emitter voltage V_{ce}.

The *input resistance* of the transistor is not a constant except at given values of V_{ce} and I_b (or V_{be}). At a chosen point on a given input characteristic, the input resistance* (h_{ie}) is given by

$$h_{ie} = \Delta V_{be}/\Delta I_b$$

where ΔV_{be} is a small change of V_{be}, and likewise for ΔI_b; h_{ie} is therefore decided by the slope of the input characteristic at a chosen point. Since h_{ie} has the dimensions of resistance, it is quoted in ohm. For example, the line LM (Figure 4.4) is a tangent to the curve at the point for which $V_{ce} = -2.0$ V and $I_b = -50$ μA and could be chosen to evaluate h_{ie}.

In specifying the parameter h_{ie}, the subscript i denotes input and the subscript e signifies common-emitter connection. The value of

---

* The term should strictly be the 'incremental resistance' or 'slope resistance' because the small *change* of current accompanying a small change of voltage is concerned. However, the adjective 'incremental' is not employed in common practice in this case.

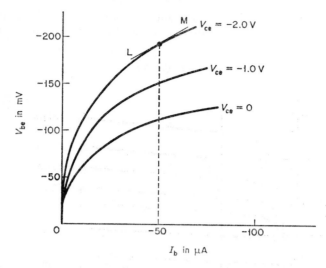

Figure 4.4. Typical common-emitter input characteristics

h_{ie} may be between 800 and 1200 Ω. It is approximately eight times the input resistance in common-base connection, which would be denoted by h_{ib}.

The *output characteristics* (Figure 4.5) of the transistor in common-emitter connection are of the collector (output) current I_c plotted against the collector-emitter (output) voltage V_{ce} for a number of values of the base current I_b.

The slope of the output characteristic, say at the reference point Q in Figure 4.5 is denoted by h_{oe}, where the subscript o denotes 'output'. Thus

$$h_{oe} = \Delta I_c / \Delta V_{ce} \quad \text{at constant } I_b$$

This parameter has the dimensions of conductance; it is the output conductance of which the reciprocal is the *output resistance*. The output resistance of the transistor is very high, of the order of 100 kΩ.

The relatively low input resistance due to the forward bias across the input junction (J_{eb}) compared with the high output resistance due to the reverse bias across the output junction (J_{cb}) gives rise to the term 'transfer-resistor' or 'transistor'.

The most important parameter of the transistor is concerned with the current gain. Denoted by h_{fe} where the subscript f denotes 'forward', it is called the *forward current transfer ratio* (or current gain). It must be ensured in specifying this current gain that the collector-emitter voltage V_{ce} remains constant. In practice, therefore, it is asserted that the current gain is with the output short-circuited to alternating current.

$$h_{fe} = \Delta I_c / \Delta I_b$$

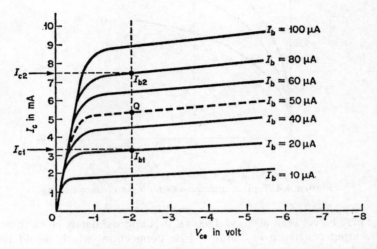

Figure 4.5. Typical output characteristics of a p-n-p germanium transistor in common-emitter connection

Referring to the linear portions of the characteristics of Figure 4.5, it is seen that

$$h_{fe} = \frac{(I_{c2} - I_{c1})}{(I_{b2} - I_{b1})} \quad \text{at a constant value of } V_{ce}$$

The *transfer characteristic* (Figure 4.6) is obtained by plotting the collector current I_c against the base current I_b for a given value of the collector-emitter voltage V_{ce}, say at -2 V.

The transfer characteristic is a straight line. Hence

$$I_c = \text{constant} \times I_b$$

This constant is denoted by β or h_{FE}, where

$$h_{FE} = I_c / I_b$$

Note the similarity and at the same time the important difference between h_{fe} and h_{FE}. The former is at a given operating point and is concerned with small changes of I_c and I_b. The latter—distinguished by the use of capital letter subscripts—is a constant obtained from a linear graph and is called the *static value of the forward current transfer ratio* or, more simply, the *direct current gain*. It is the ratio between the continuous output current (in this case I_c) and the continuous input current I_b.

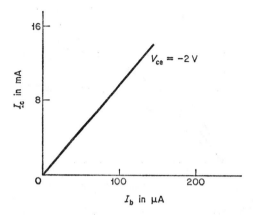

Figure 4.6. Typical transfer characteristics for a p-n-p germanium transistor in common-emitter connection

In the experiment, the values of h_{fe} and h_{FE} should be compared. They are almost equal and in future in this text will be thought of as equal. For the ACY21 they may have a value of about 120.

4.3 Characteristics of a p-n-p Transistor in Common-base Connection

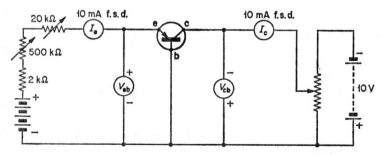

Figure 4.7. Circuit for obtaining the characteristic of a p-n-p transistor in common-base connection

The circuit of Figure 4.7 enables the characteristics to be obtained of a p-n-p transistor in common-base connection. Again, the polarities of the applied voltages should be carefully checked before the circuit is connected.

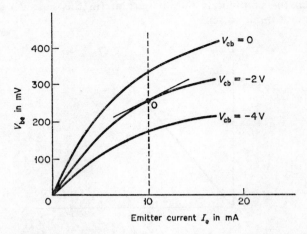

Figure 4.8. Typical input characteristics of a p-n-p transistor in common-base connection

The input characteristics (Figure 4.8) show the relationship between the base-emitter potential difference V_{be} and the emitter current I_e for a number of collector-base voltages, V_{cb}.

The *input resistance*, denoted by h_{ib}, where the subscript i denotes 'input' and b denotes 'base', is not constant but will depend on the chosen value of the emitter current I_e and the collector-base voltage V_{cb}. It is given by

$$h_{ib} = \Delta V_{be}/\Delta I_e \quad \text{at constant } V_{cb}$$

and is obtained from the slope of the tangent at the operating point on the curve. If ΔV_{be} is in volt and ΔI_e in ampere, the input resistance h_{ib} is given in ohm.

The value of the h_{ib} is between 50 and 100 ohm. This low input resistance is a consequence of the forward biased emitter-base junction and is, in many circumstances, a disadvantage of junction transistors.

The output characteristics can be obtained by setting the emitter current (I_e) at some fixed value (say 1 mA) and measuring the collector current (I_c) for various collector-base voltages (V_{cb}) between, say, 0 and −9 V. The family of output curves (Figure 4.9) shows that the output current (I_c) varies only very slightly with the collector-base voltage (V_{cb}), the characteristics being straight lines almost parallel to the V_{cb} axis. The slope of these lines gives the output conductance, denoted by h_{ob}, where the reciprocal is the output resistance of the transistor. Hence

$$h_{ob} = \Delta I_c / \Delta V_{cb} \quad \text{at constant emitter current.}$$

The output resistance is very high, certainly greater than 0.5 MΩ and is explained by the fact that the collector-base junction is reverse-biased.

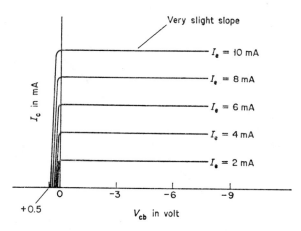

Figure 4.9. Typical output characteristics of a p-n-p transistor in common-base connection

In common-base connection the transistor acts as a constant current generator. This is because the output current (I_c) varies only slightly with the collector-base voltage (V_{cb}). Hence if a load resistance is inserted in series with the supply voltage and the collector, although this will reduce the voltage V_{cb} because of the p.d. produced across this resistance, the current through the resistance will remain almost constant even when the value of this resistance is altered.

The forward current transfer ratio, h_{fb}, also known as the forward current gain, for the transistor in common base connection can be obtained—although not very accurately—from the curves of Figure 4.9.

$$h_{fb} = \frac{\Delta I_c}{\Delta I_e} = \frac{I_{c2} - I_{c1}}{I_{e2} - I_{e1}} \quad \text{at constant } V_{cb}$$

where ΔI_c is the change of the collector current I_c for a small change ΔI_e of the emitter current I_e. The value of h_{fb} is approximately 0.99.

The *current transfer characteristic* for the transistor in common-base connection is the graph of the collector current (I_c) plotted against the emitter current (I_e) for a given value of the collector-base voltage (V_{cb}). For example, V_{cb} is set at -3.0 V and the emitter current is set at various values between 0 and 6 mA, and the corresponding collector currents are recorded. This characteristic for the

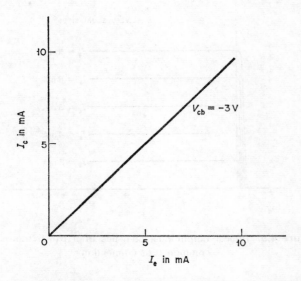

Figure 4.10. Typical current transfer characteristic for a p-n-p transistor in common-base connection

ACY21 is shown in Figure 4.10. It is typically a straight line, showing that

$$I_c = \text{constant } I_e$$

the constant in this case being denoted by α, so that

$$\alpha = I_c/I_e \quad \text{at constant } V_{cb}$$

The constant α is also the static value of the forward current transfer ratio or d.c. current gain, written as h_{FB}, the capital letter subscripts denoting the d.c. case. An experimental determination of h_{fb} and h_{FB} shows that they are virtually the same.

Table 4.1 compares the features of a p-n-p transistor in common-emitter (C.E.) and common-base (C.B.) connections.

	C.E.	C.B.
Input resistance	$h_{ie} = 1000\ \Omega$	$h_{ib} = 100\ \Omega$
Output resistance	$1/h_{oe} = 100\ \text{k}\Omega$	$1/h_{ob} = 0.5\text{M}\ \Omega$
Current transfer ratio	$h_{fe} = 120$	$h_{fb} = 0.99$

Table 4.1

Comparative features of a p-n-p transistor
in C.E. and C.B. connections
(approximate values are given)

4.4 The Relationship Between α and β

As explained in section 4.3, α or h_{FB} is I_c/I_e in the common base connection case, whereas β or h_{FE} is I_c/I_b in the common emitter case (section 4.2). To obtain a relationship between α and β consider the conventional currents in a p-n-p junction transistor (Figure 4.11).

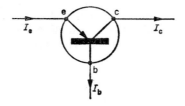

Figure 4.11. Conventional currents in a p-n-p junction transistor

From Kirchhoff's first law it is seen that

$$I_e = I_c + I_b \tag{4.1}$$

Substituting $I_e = I_c/\alpha$,

$$I_c/\alpha = I_c + I_b$$
$$1/\alpha = 1 + I_b/I_c$$
$$= 1 + 1/\beta$$

Hence
$$\beta = \alpha/(1 - \alpha) \tag{4.2}$$

and
$$\alpha = \beta/(1 + \beta) \tag{4.3}$$

These equations can equally well be written as

$$h_{FE} = h_{FB}/(1 - h_{FB}) \tag{4.4}$$

$$h_{FB} = h_{FE}/(1 + h_{FE}) \tag{4.5}$$

The value of h_{FB} is typically 0.99 (section 4.3); hence from equation (4.2),

$$\beta = h_{FE} = \frac{0.99}{1 - 0.99} = 100 \quad \text{approx.}$$

4.5 Some Features of the h (hybrid) Parameters

The parameters of the transistor so far determined from the input and output characteristics are termed *hybrid* or h parameters because they are not all alike dimensionally. Thus, h_{ie} and h_{fb} have the dimensions of resistance, h_{oe} and h_{ob} have the dimensions of conductance and the remaining ones, h_{fe} and h_{fb} are dimensionless.

To recapitulate briefly, the subscripts employed denote the following: the first letters i, o and f denote respectively input, output and forward current; the second letters e and b denote respectively common-emitter and common-base connections.

These hybrid parameters are concerned with small variations of the voltage or current about a specific operating point. They are of importance when small signal variations (the special case being small signal a.c. behaviour) are applied to the transistor. They are easy to measure and of value in enabling the performance of a transistor amplifier to be calculated. The correct terms involved are:

(*a*) In common emitter connection,

$$h_{ie} = \left(\frac{\Delta V_{be}}{\Delta I_b}\right)_{V_{ce}}$$

which is the input resistance with the output short-circuited to a.c. so that V_{ce} remains constant;

$$h_{oe} = \left(\frac{\Delta I_c}{\Delta V_{cb}}\right)_{I_b}$$

the output conductance with the input open-circuited to a.c. so that I_b remains constant;

$$h_{fe} = \left(\frac{\Delta I_c}{\Delta I_b}\right)_{V_{ce}}$$

the forward current transfer ratio (or current gain) with the output short-circuited to a.c. so that V_{ce} remains constant.

(*b*) For common base connection,

$$h_{ib} = \left(\frac{\Delta V_{be}}{\Delta I_e}\right)_{V_{ce}} \quad h_{cb} = \left(\frac{\Delta I_c}{\Delta V_{cb}}\right)_{I_e} \quad h_{fb} = \left(\frac{\Delta I_c}{\Delta I_e}\right)_{V_{cb}}$$

The d.c. biasing of a transistor is considered later (section 4.11). These hybrid parameters are not concerned with the biasing; however, they *are* concerned with small signal variations about the operating points determined by the bias. For such small signal variations, linear behaviour of the transistor about the operating point is assumed.

Manufacturers usually publish the characteristic curves of their junction transistors and quote values of h_{fe} at a specific value of I_c, say 1 mA. In general, between transistors of the same type, there is a large spread in the values of a particular parameter. For example, for an ACY21, the value of h_{fe} may have three values quoted (to indicate the spread) for $I_c = 1$ mA. These may be a minimum value of 90, a maximum of 250 and a typical value of 150.

The output resistances in both CE and CB connections ($1/h_{oe}$ and $1/h_{ob}$ respectively) are very high. In the simple basic amplifier circuit, an output load resistance R_L is connected in series between the voltage supply and the collector. The object is to obtain a voltage variation (signal) across this load which is an amplified replica of the input signal. This load resistance between collector and the battery supply terminal is in parallel with the output resistance of the transistor as regards any signal variation. This is because the internal resistance of the battery supply is negligibly small. Hence, in any analysis, the output resistance of the transistor may be considered to be infinite provided that the load resistance R_L is not too high. In practice this means that R_L should not be more than 10 per cent of $1/h_{oe}$ or $1/h_{ob}$.

The transistor is then behaving as a constant current generator working into a load resistance R_L.

4.6 Amplification in Common–Emitter Connection

To show how a junction transistor in common-emitter connection is able to amplify a small voltage change applied at its input, consider Figure 4.12. That this circuit can amplify is due to the fact that a small change ΔI_b of the base current I_b can produce a large change ΔI_c in the collector current I_c, as shown in section 4.2, because h_{fe} is about 120 (Figure 4.12).

The emitter—which is common to both the input and the output circuits—is considered to be at zero potential: it is often earthed, although this is by no means essential. As the transistor is of the p-n-p type, its collector potential must be negative with respect to earth. The supply voltage to the collector is therefore from a battery supplying a maximum of -9 V. However, it is of no use in designing an amplifier to join the collector directly to the supply voltage because it is essential for the collector potential to vary in sympathy with input voltage changes. Therefore, a load resistance R_L is connected in series between the −9 V terminal of the supply and the collector. It is across R_L that the amplified version of the input voltage change is to be produced. If R_L is 1 kΩ, it will have little effect on the collector current because the output resistance of the transistor is very high.

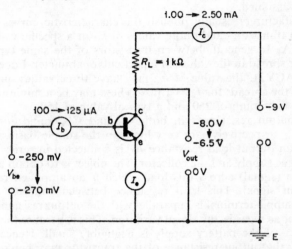

Figure 4.12. Behaviour of a voltage amplifier based upon a p-n-p transistor in common-emitter connection

To avoid, for the time being, the consideration of bias supplies to the input, suppose the base-emitter voltage V_{be} changes from −250 mV to −270 mV. As the input voltage becomes more negative, the base current increases. This increase might be from 100 μA to 125 μA if, say, an ACY21 is used. As a result the collector current I_c increases from, say, 1.0 mA to 2.5 mA. When the collector current is 1.0 mA the p.d. across R_L of 1 kΩ is $1.0 \times 10^{-3} \times 10^3 = 1.0$ V. As the battery supply e.m.f. is −9 V, the collector potential will therefore be −8.0 V. When the input potential V_{be} is altered to −270 mV (the change of V_{be} is −20 mV), the collector current increases to 2.5 mA. The p.d. across R_L therefore increases to 2.5 V and so the collector potential becomes −6.5 V.

Consequently, a change of voltage of 1.5 V is produced across the load resistance R_L for a change of the input voltage by 20 mV. This corresponds to a voltage amplification ratio (voltage gain) of $1.5/0.02 = 75$. Several features of this simple amplifier are of interest:

(a) If the input voltage is increased negatively, the collector voltage becomes more positive, and *vice versa*, if the input is increased positively, the collector voltage becomes more negative. Hence if the input voltage is alternating, the output voltage across the collector-emitter is 180° out of phase with the input voltage. The amplifier thus introduces a phase change of π or 180°.

(b) The input resistance of the amplifier is defined simply as the change in input voltage, ΔV_{be}, divided by the resulting change in input current, i.e. in the present example,

$$\frac{\Delta V_{be}}{\Delta I_b} = \frac{20 \text{ mV}}{25 \text{ μA}} = 800 \text{ Ω}$$

(c) The current gain, denoted by A_I, is defined by

$$\frac{\text{change in collector current}}{\text{change in base current}} = \frac{\Delta I_c}{\Delta I_b}$$

$$= \frac{1.5 \text{ mA}}{25 \text{ μA}} = 60$$

(d) The voltage gain, denoted by A_V, is defined by

$$\frac{\Delta V_{out}}{\Delta V_{in}} = \frac{\Delta V_{cb}}{\Delta V_{be}} = \frac{1.5 \text{ V}}{20 \text{ mV}} = 75$$

(e) The power gain is defined as

$$\frac{\text{change in output power}}{\text{change in input power}} = A_V \times A_I,$$

which, in the present example is $75 \times 60 = 4500$.

4.7 A.C. Amplification and the Load Line: A Graphical Analysis

The basic circuit of an alternating voltage amplifier using a p-n-p transistor in common emitter connection (Figure 4.13(a)) operates with a steady bias of -250 mV on the base relative to the emitter. The alternating input voltage is applied *via* a capacitance C_1 of low reactance at the frequency concerned across the base and emitter. This capacitance is necessary to isolate the a.c. input from the d.c. bias supply.

The output voltage is that occurring between the collector and the emitter. To ensure that only the alternating component of this output appears across the output terminals, a second capacitance C_2 is inserted as shown.

The alternating input signal amplitude is only 30 mV; it is a small signal input of value considerably less than the operating d.c. input bias used. This d.c. bias of -250 mV on the base with respect to the

112   *An Introduction to Semiconductors*

emitter decides the operating point Q on the I_c against V_{ce} characteristics for a mean base current, taken to be 30 μA in the present case (Figure 4.13(b)).

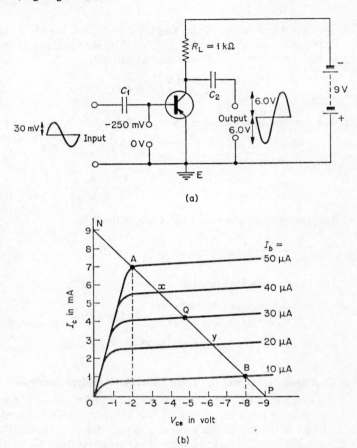

Figure 4.13. (a) An alternating voltage amplifier using a p-n-p transistor in common-emitter connection and (b) the load line drawn on the $I_c - V_{ce}$ characteristics

The applied alternating signal causes the base voltage to vary sinusoidally between −220 mV and −280 mV. The corresponding base current variations are taken to be between 50 μA and 10 μA.

A load-line AQB is drawn as a straight line through the operating point Q across the $I_c - V_{ce}$ characteristics. This load-line represents the linear relationship between current and voltage for the particular load resistance R_L used.

Suppose, for example, that R_L is 1 kΩ. To construct the load-line, consider that when V_{ce} is equal to zero, all the applied voltage must appear across R_L, i.e. the voltage drop across R_L is 9 V, the e.m.f. of the battery supply. For 9 V across 1 kΩ, the collector current must be 9 mA. Hence point N on the load-line must have co-ordinates $V_{ce} = 0$ and $I_c = 9$ mA.

When I_c is equal to zero, all the applied voltage must appear across the transistor, that across R_L being zero. Hence $V_{ce} = 9$ V and $I_c = 0$ are the co-ordinates of point P. The load-line is thus the straight line which joins N and P.

The main purpose of this load-line is to indicate the limits between which the amplifier can be operated without distortion, i.e. so that the waveform of the output signal is a true replica of that of the input signal. This freedom from distortion will be preserved provided that the sections Ax, xQ, Qy and yB on the load-line are equal in length, where the successive $I_c - V_{ce}$ characteristics drawn are for equal increments of the base current I_b.

With the operating point chosen at Q and with the input signal causing the base current to vary between 50 and 10 μA, the load-line indicates that V_{ce} varies between −2.0 V and −8.0 V. Hence an output signal voltage of 6.0 V peak-to-peak is obtained.

Distortion of the output signal will occur if the variation in the base current takes this signal beyond the linear region of operation defined by the points A and B. As the voltage across the load approaches 9 V, if the base current increases further, the collector current cannot follow. This condition occurs as the operating point moves beyond A. The transistor is then said to be *saturated* or *bottomed*. Beyond the point B on the load-line, when the collector-current is very small, distortion will occur as the transistor approaches *cut-off*.

This graphical approach shows that the amplifier will have the following features:

(a) Current gain $A_I = \Delta I_c/\Delta I_b = 6$ mA/40 μA $= 150$.

(b) Voltage gain $A_V = \Delta V_{ce}/\Delta V_{be} = 6.0$ V/60 mV $= 100$.

(c) Power gain $A^I V_V = 150 \times 100 = 1.5 \times 10^4$

4.8 Circuit Analysis by Means of a Simple Equivalent Circuit

The alternating current behaviour of any transistor amplifier can be calculated by use of the hybrid parameters together with a circuit which is a simple model of the amplifier. Dealing with common emitter connection, the parameters which will determine the behaviour of the amplifier are the input resistance h_{ie}, the output resistance $1/h_{oe}$ and the forward current transfer ratio h_{fe}. If the output resis-

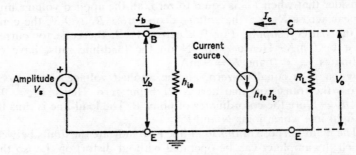

Figure 4.14. Simple equivalent circuit of a common-emitter amplifier

tance is high (at least ten times the load resistance) the amplifier can be represented by the equivalent circuit of Figure 4.14.

The equivalent circuit is derived on the basis that a sinusoidal input signal of amplitude V_s is applied between the base and the emitter and causes a base current of I_b to flow through the input resistance h_{ie}. The output circuit represents the behaviour of the transistor as a source of constant current $h_{fe} I_b$ (i.e. the forward current transfer ratio times the base current) which flows through the load resistance R_L to produce an output voltage of V_o.

Example 4.8

A p-n-p junction transistor is used as an audio-frequency voltage amplifier in common-emitter connection. The load resistance is 1 kΩ. If the transistor parameters are $h_{ie} = 1100 \; \Omega$, $h_{fe} = 50$ *and* $h_{oe} = 2.5 \times 10^{-5}$ *siemen, calculate the voltage gain and the power gain of the amplifier.*

As the output conductance $h_{oe} = 2.5 \times 10^{-5}$ siemen, the output resistance is $1/(2.5 \times 10^{-5}) = 4 \times 10^4 \; \Omega$. The load resistance of 1 kΩ is in parallel with this output resistance. As its value is less than 10 per cent of the output resistance, the equivalent circuit of Figure 4.14 is valid. The voltage gain is given by

$$A_V = \text{voltage out/voltage in}$$

Voltage out $= h_{fe} I_b R_L$, whereas voltage in $= I_b h_{ie}$. Hence

$$A_V = \frac{h_{fe} I_b R_L}{I_b h_{ie}} = \frac{h_{fe} R_L}{h_{ie}} = \frac{50 \times 1000}{1100} = 45.5$$

The current gain is given by

$$A_I = I_{out}/I_{in} = h_{fe} = 50$$

and

$$\text{power gain} = \frac{\text{power out}}{\text{power in}}$$
$$= \frac{I_{\text{out}} \times \text{voltage out}}{I_{\text{in}} \times \text{voltage in}}$$
$$= \text{current gain} \times \text{voltage gain}$$
$$= A_\text{I} \times A_\text{V} = 2275$$

A power gain of 2275 is equivalent to 33.6 dB (see section 4.9).

4.9 Logarithmic Units for Power Ratios: the Decibel

It is convenient to express power, voltage and current ratios on a logarithmic scale, known as the decibel scale. The ratio of two powers P_1 and P_2 expressed in bels is given by $\log_{10}(P_2/P_1)$. In decibels (dB) this is

$$10 \log_{10}(P_2/P_1) \tag{4.6}$$

Very often P_1 is chosen to be some reference power level.

For an amplifier, P_2 would be the output power and P_1 the input power. If the input and output resistances of an amplifier are both equal to R, $P_2 = V_2^2/R$ and $P_1 = V_1^2/R$, where V_2 is the output voltage and V_1 is the input voltage. Under these circumstances, (4.6) can be written

$$10 \log_{10}[(V_2^2/R)/(V_1^2/R)] = 20 \log_{10}(V_2/V_1)$$
$$= 20 \log_{10} A_\text{V} \tag{4.7}$$

Expressed in terms of the input current I_1 and the output current I_2, (4.6) becomes

$$10 \log_{10}[(I_2^2 R)/(I_1^2 R)] = 20 \log_{10}(I_2/I_1)$$
$$= 20 \log_{10} A_\text{I} \tag{4.8}$$

It is common practice—even though it is somewhat misleading—o ignore the fact that the output and input resistances are generally not equal, and simply to express the voltage gain of an amplifier in dB by $20 \log_{10}(V_o/V_1)$ where V_o is the output voltage and V_1 is the input voltage.

When a number of amplifier stages each of known gains are coupled together the overall gain is the product of the individual gains. If these gains are expressed on a logarithmic scale, the overall gain is simply the sum of the individual gains.

A number of voltage and power ratios are expressed on the dB scale in Table 4.2.

A negative value for the gain expressed in dB means that the output power is below the reference level or that the output voltage

Voltage ratio	Power ratio	dB
1	1	0
1.414	1.995*	3
1.995*	3.98†	6
3.162	10	10
10	100	20
31.62	1000	30
100	10^4	40
1000	10^6	60

* Approximately 2; † approximately 4.

Table 4.2

Voltage and power ratios expressed on the dB scale

of an amplifier is less than the input voltage. For example, if the output voltage of the amplifier is 0.1 of the input voltage, dB = $20 \log_{10}(0.1) = 20 \times \bar{1} = -20$.

An amplifier is usually designed to have a uniform voltage gain $A_V (= V_{\text{out}}/V_{\text{in}})$ over a wide frequency range. Such an amplifier is said to have a *flat response*. The band width of an amplifier is defined as $(f_2 - f_1)$ where f_2 is the upper frequency at which the power gain has dropped to one-half of its mid-band value, i.e. the power gain is '3 dB down' and the voltage gain has dropped to $A_V/\sqrt{2}$. Similarly f_1 is the lower frequency at which the voltage gain has dropped to $A_V/\sqrt{2}$.

4.10 Leakage Current and Thermal Runaway

For a junction transistor in common-base connection, if the emitter is open-circuited and the collector-base junction is reverse biased, a leakage current, denoted by I_{CBO}, flows between the collector and the base. For a low-power germanium transistor, I_{CBO} may have a value of 3.0 μA at 20°C but is considerably smaller for a silicon transistor.

In common-emitter configuration with the emitter-base junction forward biased and the collector-base junction reversed biased, if the base is disconnected, the leakage current in the collector circuit, I_{CEO}, is much larger and may be 100 μA for a germanium transistor at 20°C. This leakage current doubles approximately for every 10°C rise in temperature. A silicon transistor of similar power handling capability may have $I_{\text{CEO}} = 10$ μA at 20°C. Manufacturers usually quote the value of I_{CBO} at 20°C.

The value of I_{CEO} may be calculated if I_{CBO} is given. Thus the leakage current I_{CBO} continues to flow from the base to the collector. As the base current is zero, there must be an emitter to base current of I_{CBO} (Figure 4.15). The transistor action produces a current of

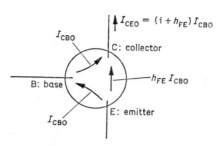

Figure 4.15. The relationship between the leakage current I_{CBO} and I_{CEO}

$h_{fE} I_{CBO}$ in the collector circuit. Hence the total leakage current in the collector circuit is

$$I_{CEO} = (1 + h_{fE})I_{CBO} \quad \text{or} \quad I_{CEO} \simeq h_{FE} I_{CBO} \qquad (4.9)$$

The effect of leakage current is two-fold:

(i) A rise in temperature will increase the leakage current and shift the operating point, so causing distortion of the amplified signal.

(ii) The leakage current causes a temperature rise which results in increased leakage current and further temperature rise. This cumulative process can result in the destruction of the transistor; it is called *thermal runaway*, and is much more likely to occur in germanium than in silicon transistors. The shift of the operating point due to leakage current is illustrated in Example 4.10.

Example 4.10

The values of the leakage current I_{CEO} in a germanium transistor at various temperatures are listed below:

Temperature (°C)	20	30	40	50	60
Leakage current, I_{CEO} μA	60	120	240	480	960

This transistor has $h_{FE} = 50$. It is used as an amplifier in common-emitter connection with a d.c. bias current of 40 μA. Calculate the effect on the collector current I_c as the junction temperature rises from 20°C to 60°C.

At 20°C,
$$I_c = h_{FE} I_b + I_{CEO}$$
$$= (50 \times 40 + 60) \text{ A } \mu = 2.06 \text{ mA}$$

This value is not very different from the ideal case when the leakage current is zero, i.e.
$$I_c = h_{FE} I_b = 50 \times 40 \text{ μA} = 2.0 \text{ mA}.$$

At 60°C,
$$I_c = h_{FE} I_b + I_{CEO}$$
$$= (50 \times 40 + 960) \text{ μA} = 2.96 \text{ mA}$$

The effect of the leakage current at 60°C is therefore to increase the collector current by 0.96 mA, i.e. by nearly 50 per cent. This very significant increase affects considerably the d.c. bias point. The operating region could move away from the linear portion of the characteristics and cause distortion.

The circuit with its d.c. biasing must be designed to prevent thermal runaway and to stabilize the working point so that it changes only marginally with temperature and does not introduce distortion.

4.11 Biasing the Junction Transistor

One of the problems in designing a transistor amplifier is to fix and maintain the collector to emitter voltage V_{ce} and the emitter current I_e, i.e. to arrange for the transistor to operate about the working point Q (Figure 4.13(b)), which is maintained despite temperature fluctuations. The biasing circuit must therefore not only stabilize the working point but also prevent thermal runaway. A number of biasing methods for a transistor in common-emitter connection are described.

(*a*) *The fixed bias circuit* A resistance R_b is connected between the supply voltage terminal and the base (Figure 4.16). The base current I_b required to establish the best working point is known from the characteristics (Figure 4.13(b)). It is then a simple matter to calculate R_b because

$$I_b = V_{cc}/R_b \qquad (4.10)$$

where V_{eb} is negligible, V_{cc} being the battery supply voltage. This circuit is thermally unstable: any increase in temperature which causes the leakage current I_{CEO} to increase will increase the collector current. Even if thermal runaway does not occur, distortion is likely at elevated temperatures.

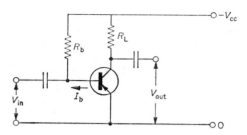

Figure 4.16. The fixed bias circuit

(*b*) *Collector-to-base bias* An improvement in stability is obtained if the resistance R_b is connected between collector and base (Figure 4.17). If the collector current tends to increase, the voltage V_{ce}

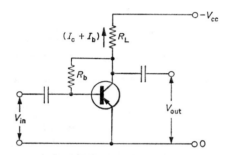

Figure 4.17. Collector-to-base bias

decreases. This is d.c. negative feedback. Hence I_b also decreases and the collector current does not increase as much as if fixed bias had been used. Neglecting the voltage V_{be}, the required value of the base resistance R_b is calculated, knowing V_{cc}, I_c, I_b and R_L, from

$$I_b = \frac{V_{cc} - I_c R_L}{R_L + R_b} \qquad (4.11)$$

One fundamental disadvantage of this circuit is that part of the alternating output signal which is 180° out of phase with the input signal is fed back into the input circuit via R_b. This phenomenon is a.c. negative feedback (see section 4.13). The d.c. negative feedback aids the stability; the a.c. feedback reduces the gain of the stage.

(*c*) *Collector-to-base bias with the base resistor decoupled* The reduction of gain due to the a.c. feedback resulting in case (*b*) may be eliminated by dividing R_b into two equal parts and connecting the junction between these resistors to the common-emitter *via* a capacitance C_1 which has a negligible reactance at the signal frequencies (Figure 4.18). The base resistance R_b is said to be decoupled at the signal frequencies. The alternating output signal is now not fed back into the input circuit.

Collector-to-base biasing is often used, especially with silicon transistors in which the leakage currents are low.

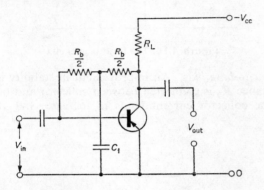

Figure 4.18. Collector-to-base bias with R_b decoupled

(*d*) *Potential divider and emitter resistor stabilizing bias* The necessary steady bias on the base is provided by a potential divider consisting of resistances R_1 and R_2 across the battery supply voltage V_{cc} (Figure 4.19). Furthermore, the emitter is not connected directly to the (often earthed) terminal of the supply—the positive terminal for a p-n-p transistor—but via a series resistance R_e. This resistance R_e carries the emitter current I_e and so there is a steady bias on the emitter relative to earth of $-I_e R_e$. When an alternating signal is applied to the input via the capacitance C, there will also be an

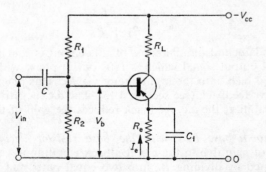

Figure 4.19. Bias with a potential divider and emitter resistor stabilizing

alternating component of the emitter current. To ensure that this does not introduce a corresponding alternating bias on the emitter, the capacitance C_1 is connected in parallel with R_e. The reactance of C_1, which is $1/(2\pi f C_1)$ where f is the input signal frequency, is chosen by virtue of the magnitude of C_1, to be much smaller in magnitude than R_e.

With this provision of R_e, any increase in collector current causes an increased voltage drop across R_e so that V_{eb} decreases, causing a smaller base current to flow. This in turn means that I_e will increase less than it would have done had there been no self-biasing resistance R_e.

By choice of R_1 and R_2, the current through them is arranged to be several times the base of current I_b. Consequently, V_{be} is virtually independent of I_b, and

$$V_{be} = V_b - I_e R_e \qquad (4.12)$$

where V_b is the voltage drop across R_2. Hence

$$I_e = V_b/R_e \quad \text{approx.} \qquad (4.13)$$

because V_{be} is very small.

The steady emitter current or the collector current ($I_e \simeq I_c$) can be calculated from equation (4.13) merely from a knowledge of the resistance values forming the potential divider and the value of R_e. This method of obtaining bias by potential divider and emitter resistor stabilizing is often used.

4.12 A Two-stage Common-emitter Transistor Amplifier

The most frequently used method of coupling together two stages in an alternating voltage amplifier is by resistance-capacitance coupling (Figure 4.20(a)). This circuit uses two p-n-p junction transistors (ACY21) with potential divider, emitter resistor stabilizing bias, and the component values given are suitable for alternating voltages of audio-frequency over the range from 20 Hz to 30 kHz.

To indicate how the alternating voltage which appears across the load resistance R_{L1} of stage 1 (which is an amplified replica of the input to the amplifier) is connected to the input of the second stage 2, consider that a capacitance C_1 and resistance R_1 are in series across R_{L1} (ignoring other components for simplicity). It is seen from Figure 4.20(b) that the alternating voltage across R_1 is $R_1/[\sqrt{R_1^2 + (1/2\pi f C_1)^2}]$ for a frequency f. If C_1 is large enough for $(1/2\pi f C_1)$ to be much smaller than R_1, almost the whole of the alternating voltage across R_{L1} appears across R_1, forming the input to stage 2.

In an experiment, the alternating input voltage V_1 can be conveniently obtained from a transistor oscillator with a potential divider across its output terminals to ensure that an input signal is obtained which is small enough not to overload the amplifier and cause distortion. This input voltage can be measured by either a multirange a.c. transistor voltmeter or a calibrated cathode-ray

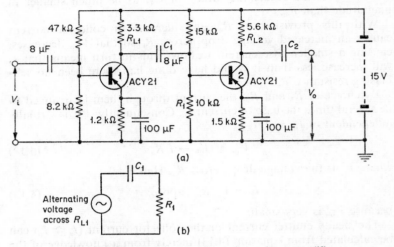

Figure 4.20. A two-stage alternating voltage amplifier

oscillograph. The output voltage V_0 from the amplifier is across R_{L2}, conveniently isolated from the steady voltage by the decoupling capacitance C_2 so that V_0 appears across the collector (except for the low reactance of C_2) of stage 2 and earth. V_0 is also measured with the multirange meter or the cathode-ray oscillograph.

If the oscillator used to provide the input is of suitably variable frequency, the overall voltage amplification A_V provided by the two-stage amplifier may be plotted against the frequency over the range, say, from 20 Hz to 30 kHz.

4.13 Feedback

Feedback in an amplifier occurs when the output exerts some influence on the input. This may occur fortuitously or it may be introduced deliberately. When the signal fed back from the output to the input is 180° out of phase (in anti-phase) with the input signal, the amplification is decreased and the feedback is said to be *negative* or *degenerative*. On the other hand, when the signal fed back is in phase with the input signal, the amplification is increased and the feedback is *positive* or *regenerative*.

It would seem at first that positive feedback is desirable. However, it is practically never used in amplifier design because its effect is cumulative leading to instability. Positive feedback increases the input signal which in turn increases the ouput signal which then increases the input still further. Though not used in amplifiers it is nevertheless employed in oscillators.

Negative feedback is always or almost always used in the design of amplifiers. Though it must clearly result in loss of amplification because part of the output is fed back to reduce the input, a number of distinct advantages accrue from its use which far outweigh this reduced amplification. In any case, further amplification can be obtained by simply adding another stage.

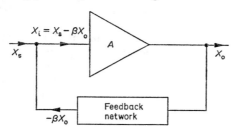

Figure 4.21. Block diagram to illustrate negative feedback

The conventional block symbol for an amplifier is a triangle (Figure 4.21) in which the letter A is inscribed to denote its gain. The signal from input to output is imagined to travel in the direction of the apex of the triangle: the input is therefore at the base; the output is at the apex.

Let the actual input signal be X_s and the output signal that this would give be X_o. Suppose that by some means a fraction β of this output is fed back to the input circuit so as to be 180° out of phase with the input, i.e. negative feedback is practised. The input signal now becomes X_1 given by

$$X_1 = X_s - \beta X_o$$

Before feedback was applied, the input was X_s and the output was $X_o = X_s A$. With negative feedback applied, the signal $-\beta X_o$, equal to $-\beta X_s A$, is introduced into the input. Hence to obtain the same output as before the input signal must be increased by $\beta X_s A$, to become $X_s(1 + \beta A)$. The gain of the amplifier with negative feedback applied is therefore A_f given by

$$A_f = \frac{X_o}{X_s(1 + \beta A)} = \frac{A}{1 + \beta A} \quad (4.14)$$

Although the fraction β (known as the *feedback factor*) may be small, A can be very large so that the term βA is large compared with unity. In this case, 1 is negligible in the denominator of equation (4.14) so that

$$A_\mathrm{f} = \frac{A}{\beta A} = \frac{1}{\beta} \tag{4.15}$$

This equation is of paramount importance in negative feedback. It shows that the gain of the amplifier with negative feedback depends almost entirely on the circuit arrangement which decides β and is largely independent of A (often called the *open-loop gain* of the amplifier, i.e. the gain with zero feedback).

In a simple case, therefore, the gain is virtually independent of the transistor used. If a transistor becomes faulty and is replaced by another one of the same type but different characteristics (and the characteristics of transistors of the same type may vary considerably because of difficulties in maintaining a consistent semiconductor manufacturing process) the amplifier circuit behaviour is unchanged.

Other benefits resulting from the use of negative feedback are:

(a) very stable operation;
(b) low distortion,
(c) reduction of noise;
(d) the ability to control the input and output resistance of the circuit.

Further discussion of these factors is beyond the scope of this text.

The remainder of this chapter is concerned with various useful simple but basic circuits amongst the very large number that can be constructed using junction transistors.

4.14 A Constant-Current Source

A source which provides a constant current to a load is frequently required in laboratory practice. A simple example is the necessity for the current through a filament lamp to be constant if the light output from the lamp is to be unvarying.

Ideally a constant current source should have an output resistance ($\Delta V/\Delta I$, where ΔV is the change of voltage for a current change of ΔI) which is very large (ideally, infinite) so that the introduction of any load resistance R_L has very little effect on the current that flows through it. Thus if R_L varies its effect is insignificant on $\Delta V/\Delta I$ (because R_L is much less than $\Delta V/\Delta I$) and so on the current I which flows through it.

A transistor in common-base connection has a very high output resistance, $1/h_\mathrm{ob} = \Delta V_\mathrm{cb}/\Delta I_\mathrm{c}$ (section 4.3). However, its emitter-base voltage must be kept constant. In the constant-current circuit (Figure 4.22), therefore, the silicon n-p-n transistor used has its emitter-base voltage supply from a Zener diode stabilized supply so that the p.d. across the 5 kΩ resistor is 11.2 V at all times because of the two Zener diodes (Type M–ZE which each stabilize at 5.6 V) in series across it. The 5 kΩ potentiometer provides a coarse control and the variable 500 Ω resistor a fine control for the current required.

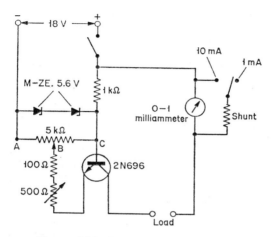

Figure 4.22. A constant-current source

Some negative d.c. feedback is present. This is because, if the collector current I_c of the n-p-n junction transistor (type 2N696) increases for some reason, the emitter current I_e also increases. Hence the p.d. across the section AB of the 5 kΩ resistor increases and so that across the section BC (which decides the emitter-base voltage on the transistor) must fall; this tends to reduce I_e. The resistance in series in the emitter circuit of the transistor also reduces the emitter-base voltage if I_e increases.

The collector-base voltage supplied to the transistor is that across the 1 kΩ resistor. It is 6.8 V in the circuit of Figure 4.22. Negligible current change (<0.5 per cent) can be detected by the milliammeter in the output circuit of the transistor as the load resistance R_L is altered, until the transistor 'bottoms' or 'saturates'. This will occur only when all the 6.8 V is dropped across the load resistance.

An experiment based on such a circuit can be undertaken to show that for a current of 10 mA through R_L, the change of this current is less than 0.5 per cent for variations of R_L between 0 and 600 Ω. If the current maintained through R_L is 1 mA, R_L can be varied from 0 to 6000 Ω with less than 0.5 per cent load current change, and with the current at 0.5 mA, R_L can be altered between 0 and 12 000 Ω.

This constant current source is of particular value in the operation of the following apparatus:

(i) To measure the variation of the conductivity of germanium with temperature (section 1.19).

(ii) To examine the characteristics of a thermistor (section 5.16). The small current passed through the thermistor produces negligible heating but enables the resistance to be measured at a number of temperatures.

(iii) To operate Hall probes in that magnetoresistance effects are eliminated.

4.15 Stabilized Power Supplies

The basic Zener diode regulator circuit (section 3.16) enables a supply of voltage to be stabilized (section 3.17) and is frequently used when the current demand is small. The limitation is the power-handling capacity of the Zener diode. In general, if the current demand exceeds 50 mA, or if increased stability is needed, or if a variable output voltage is required, a series transistor voltage regulator is used.

A p-n-p junction transistor has its base voltage maintained constant with respect to the positive terminal (often earthed) of the power pack or other supply source by means of a Zener diode (Figure 4.23). The p.d. across the resistor R_b provides the collector-base voltage. The output voltage V_o is given by

$$V_o = V_z + V_{eb} \qquad (4.16)$$

As V_{eb} is small, approximately 0.2 V for a germanium and approximately 0.5 V for a silicon transistor, $V_o \simeq V_z$, where V_z is the constant p.d. across the Zener diode.

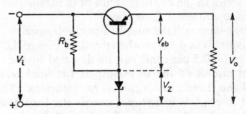

Figure 4.23. A series transistor regulator

If the input voltage V_i (usually that of a power pack) changes for any reason (e.g. a.c. mains supply fluctuation or variation of the load R_L) the collector-base voltage V_{cb} (across R_b) changes by the same amount because $V_i = V_{cb} + V_z$ and V_z is constant. The output voltage V_o remains unchanged however, because the change of V_{eb} is negligibly small.

For good regulation, the d.c. current gain h_{FE} of the transistor used should be large and the current through the Zener diode should be large compared with the base current of the transistor.

A useful circuit for operating semiconductor electronic circuits provides a constant 10 V able to supply currents of up to 40 mA (Figure 4.24). The subminiature mains transformer used with a secondary providing $12 - 0 - 12$ V is a most convenient component for use in a small power supply because it fits neatly on to a circuit

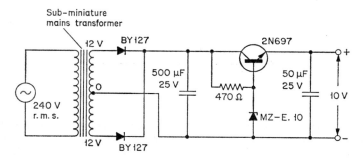

Figure 4.24. A 10 V stabilized voltage supply

board with the other components. However, its mains input terminals are not shrouded so that some thought should be given in construction to avoid the hazard of exposed 240 V terminals. Though simple to construct and providing a useful laboratory unit, this power supply has two main disadvantages:

(a) No provision exists for varying the output voltage which is almost equal to V_Z, the voltage of the reference Zener diode MZ–E 10.
(b) Changes in V_{eb} and V_Z due to changes in temperature appear at the output.

Note that the series transistor which, in effect, absorbs any voltage variations, must be capable of dissipating the power produced at its junctions. It may therefore require to be mounted on a heat sink. For the 2N697 transistor of Figure 4.24, which is within a metal can termed a T.O.5 (transistor outline) configuration, a SISTASINK type 2215 heat sink is ideal.

The power produced within the transistor is not necessarily a maximum when the current through it is a maximum. The power is most readily calculated approximately from the product of the load current I_L and the voltage V_{cb} across the reverse-biased collector-base junction (i.e. $I_L V_{cb}$).

On occasions when a large load current is demanded, a number of transistors connected in parallel serve as the series control element. In such a case, it must be ensured that each transistor carries approximately the same current.

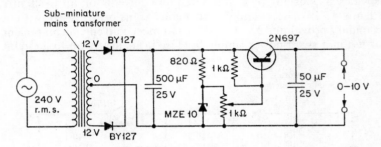

Figure 4.25. A variable stabilized voltage supply

To provide a variable stabilized output voltage, a 1 kΩ potentiometer connected across the Zener diode has its tapping point connected to the base of the series control transistor (Figure 4.25). The regulation of this variable supply is slightly inferior to that obtained with a fixed reference voltage from a Zener diode.

To obtain much improved stabilization, a difference amplifier circuit is used (Figure 4.26).

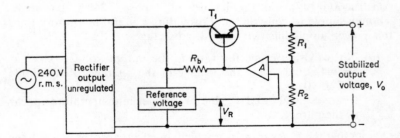

Figure 4.26. A stabilized voltage supply utilizing a difference amplifier

The transistor T_1, or a number of transistors in parallel, form the series regulating elements. All the load current passes through this regulator. As the intention is to maintain constant the output voltage V_o, there must be maintained across the series regulator the difference in voltage between the input (the unstabilized supply) and the output voltage.

The triangle represents a difference amplifier with a voltage gain of A. The two signals (one a reference voltage provided, for example, by a Zener diode and the other from a tapping on the potential divider R_1R_2 across the output) enter the base of the triangle, representing the input to the amplifier. The difference between these voltages after amplification by the factor A appears as the output at the apex of the triangle. This output controls the current in the biasing resistance R_b and hence the collector-base voltage V_{cb}.

If the output voltage V_o should rise for any reason, the input signal to the difference amplifier exceeds the reference voltage and the current through R_b increases. This increase absorbs the increase in output voltage which returns immediately to its stable level decided by the ratio R_1/R_2.

In the steady state the circuit will arrange for the difference to be negligible between the two input voltages to the difference amplifier. Hence

$$V_o = V_R + \left[\frac{R_1}{R_1 + R_2}\right] V_o \qquad (4.17)$$

where V_R is the reference voltage.

$$V_o = V_R \left[1 + \frac{R_1}{R_2}\right] \qquad (4.18)$$

It is apparent from equation (4.18) that the output voltage V_o can be varied by adjusting the ratio R_1/R_2, the minimum voltage being V_R.

Figure 4.27 shows a practical circuit of this form of stabilized supply. The difference amplifier is a silicon n-p-n transistor type 2N930. One input to this transistor is the reference voltage across the Zener diode; the other is from the potential divider across the output.

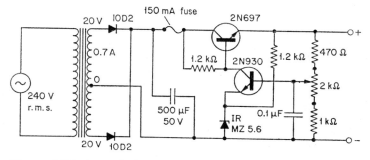

Figure 4.27. A practical stabilized voltage power-pack based on a difference amplifier

4.16 Transistor Oscillators

An oscillator circuit is basically that of an amplifier in which a fraction of the output power is fed back to the input circuit so as to be in phase with the initiating input signal. This positive or regenerative feedback provides a self-generating current or voltage variation at a frequency dependent upon the component values used.

An RC Sinusoidal Oscillator

The output from a p-n-p transistor (e.g. type ACY21) in common-emitter connection forming stage 1 is resistance-capacitance (RC) coupled (*cf.* section 4.12) to the input of a second similar transistor forming stage 2 (Figure 4.28). As each stage introduces a phase change of 180° (section 4.6) the resultant phase shift introduced by the two coupled transistors is 360°.

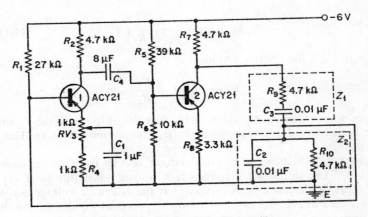

Figure 4.28. An *RC* sinusoidal oscillator

The load in the output of the second stage consists of two impedances in series: Z_1 comprising R_9 and C_3 in series and Z_2 comprising C_2 and R_{10} in parallel. The alternating voltage across Z_2 is made the input to the first-stage transistor. This fed-back input will be in phase with the output and so also in phase with the existing input provided that the alternating voltage across Z_2 is in phase with that across Z_1 and Z_2 in series. The feedback will then be positive so the circuit will oscillate. It can be shown that this zero phase shift occurs at a frequency f decided by

$$2\pi f = 1/\sqrt{(C_2 C_3 R_9 R_{10})} \qquad (4.19)$$

which is the frequency of oscillation.

It is usual to make $C_2 = C_3$ and $R_9 = R_{10}$ so that equation (4.19) becomes
$$f = 1/2\pi C_3 R_9$$
With $C_3 = 0.01$ μF and $R_9 = 4.7$ kΩ, as in Figure 4.28,
$$f = 1/(2\pi \times 10^{-8} \times 4.7 \times 10^3) = 3300 \text{ Hz} \quad \text{approx.}$$
The output across resistance R_8 is of particularly good sinusoidal waveform.

A Crystal-controlled Oscillator

If a slice of a piezoelectric crystal—usually quartz—has electrodes plated on opposite faces and a potential difference is applied between these electrodes, the crystal slice will become physically strained and a small change in its physical dimensions will occur. Conversely, if the slice is strained it becomes electrically polarized and free electric charges appear on the plated faces.

If the slice is correctly cut with respect to the crystal axes, correctly mounted and made to vibrate at its natural frequency by the application of a specific resonant alternating p.d., a very stable frequency source is obtained. Slices of quartz are cut to dimensions for which the natural frequencies are within the range from a few kHz to a few MHz. A quartz crystal slice with a frequency of 465 kHz is common and readily available. Such a crystal is used in the oscillator shown in Figure 4.29.

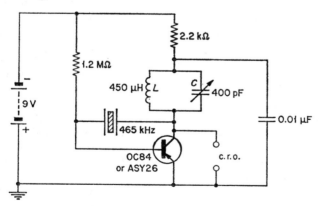

Figure 4.29. A crystal-controlled oscillator

The p-n-p transistor (type OC84 or ASY26) is operated with its emitter earthed and an inductance-capacitance (*LC*) circuit with a resistor of value 2.2 kΩ as a load in series between the —9 V terminal of the supply and the collector. The collector is coupled to the base of the transistor by the capacitance of the quartz crystal between its plates. The coupling provides positive feedback between the collector and the base. The 1.2 MΩ resistor is provided to ensure the correct d.c. base bias; the 2.2 kΩ resistor ensures the appropriate collector d.c. potential.

The tuned circuit LC does not determine the frequency of oscillation; it serves to vary (on altering the capacitance C) the amplitude of the oscillations which will be a maximum when the resonant frequency $1/[2\pi\sqrt{(LC)}]$ of this circuit and that of the crystal are the same.

The waveform of the oscillating voltage output is observed by means of a cathode ray oscillograph joined between the collector and earth. This will be the same as that across the LC circuit because one end of this circuit is earthed via the 0.01 μF capacitor as regards a.c. and the other is joined to the collector.

A Phase-shift Oscillator

Consider a capacitance C and a resistance R in series. The alternating current through this combination leads in phase with the alternating voltage across it by an angle α given by

$$\tan \alpha = 1/\omega CR$$

so that
$$\alpha = \tan^{-1}(1/\omega CR),$$

where $\omega = 2\pi f$, f being the frequency of the alternating current.

At a particular frequency f, it is clearly possible to choose the values of C and R to be such that $\alpha = 60°$.* The use of three such stages in cascade can hence be used to provide an overall phase shift of 180°. If such an arrangement is used in conjunction with a transistor in common-emitter connection, it is possible for the 180° phase change introduced by the transistor itself (section 4.6) to be added to the 180° obtained by the use of the three RC stages to arrange that the output signal is fed back to produce an in-phase input signal so that an oscillator is obtained.

The circuit of a phase-shift oscillator (Figure 4.30) makes use of a high gain n-p-n transistor (for example, type BC107). The three phase shifting RC circuits consist of C_1R_1, C_2R_2 and C_3 in conjunction with the input resistance of the transistor, where $C_1 = C_2 = C_3 = 0.01$ μF and $R_1 = R_2 = 10$ kΩ. With a 4.5 V supply, a resistor of 180 kΩ arranges an appropriate positive bias on the base of the transistor which has its emitter connected to the earthed line. A load resistance of 5.6 kΩ is in series with the collector and the

---

* A convenient value for explanation; in fact, the three RC sections do not have to be identical and rarely are. Indeed, for other RC networks, the frequency will adjust itself so that the part of the signal providing a phase shift of 180° will be further amplified in preference to any other.

positive terminal of the supply. The voltage at the collector will vary in antiphase with any voltage variation on the base relative to the emitter. The three *CR* circuits introduce a further 180° phase change so that any voltage variation on the base is supplemented by an in-phase positive feedback.

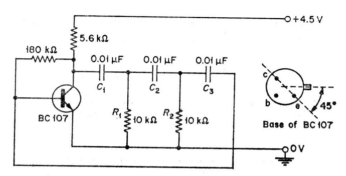

Figure 4.30. A phase-shift oscillator which makes use of an n-p-n transistor

A cathode-ray oscillograph connected between the collector and the emitter enables the waveform (which is sinusoidal) of the oscillating voltage output to be examined and its frequency to be determined.

The Multivibrator

A circuit of particular interest which is often used as a pulse generator is the multivibrator or *astable* circuit. The term 'astable' means 'not stable' indicating that oscillations are obtained, although in this case the waveform of the oscillations is non-sinusoidal. Indeed an output of rectangular waveform is produced.

A voltage (or current) of rectangular waveform can be shown to be equivalent to the summation of a sinusoidal alternating voltage of a fundamental frequency *f* plus a very large number of components of frequencies $2f, 3f, \ldots, nf$, where *n* is an integer, which are the second, third and, in general, the *n*th harmonics respectively.

A free-running multivibrator, Figure 4.31(a), is so-called because it will begin to oscillate (generating an approximately rectangular waveform output) as soon as the supply is switched on. The multivibrator has two apparently stable (quasi-stable) states and makes a periodic transition from one state to the other. One state exists with transistor T_1 (Figure 4.31(a)) fully conducting and transistor T_2 cut off while the other state is the reverse, i.e. T_2 fully conducting and T_1 cut off.

Using p-n-p germanium transistors (e.g. ACY21) with an 8 V power supply, when T_1 is cut-off this voltage will exist between point A and earth, so $(V_{ce})_{T1} = 8$ V. When T_2 is fully conducting almost all the 8 V is dropped across the resistance R_1 and $(V_{ce})_{T1} = 0$. Hence the potential at A switches between -8 V and 0 while that at

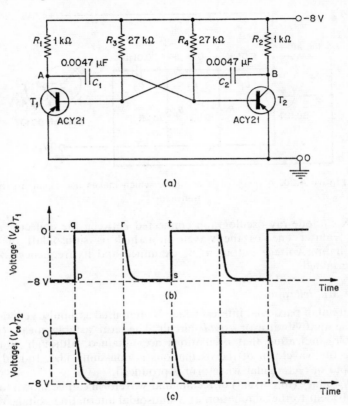

Figure 4.31. A free-running multivibrator utilizing two p-n-p transistors

B switches between 0 and -8 V. Each of these states can be observed with an oscilloscope. One waveform displayed is that of $(V_{ce})_{T1}$ between A and earth (Figure 4.31(b)) while the other is $(V_{ce})_{T2}$ between B and earth (Fig. 4.31(c)).

Any change in the voltage occurring at A is transmitted via capacitance C_1 to the base of T_2; correspondingly, a change in voltage at B is transmitted via capacitance C_2 to the base of T_1.

No two transistors, even if of the same type, behave identically. Assume, therefore, that when the supply voltage is first switched on T_1 conducts more than T_2. The potential at A goes from -8 V to zero and this transmits a positive pulse to the base of T_2 which reverse biases its emitter-base junction and cuts off T_2. The potential at B therefore drops to -8 V. This transmits a negative pulse to the base of T_1 via C_2 and turns it full on. While this seemingly steady state exists, C_1 begins to discharge through the resistance R_4 so that the base of T_2 becomes more negative with respect to the emitter until the emitter-base of T_2 is forward biased and starts to conduct. The resulting voltage change at B is transmitted via C_2 and serves to turn off T_1, while the resulting voltage at A ensures that T_2 is fully conducting. This transition from one state to another occurs with a period T given approximately by

$$T = (CR/0.77) \text{ s}$$

where C is in farad and R is in ohm.

An experiment based on the circuit of Figure 4.31 conveniently utilizes ACY21 transistors, $C_1 = C_2 = 0.0047\,\mu\text{F}$, $R_3 = R_4 = 27\,\text{k}\Omega$ and the load resistances R_1 and R_2 each equal to 1 kΩ. The waveform of the voltage variation V_{ce} between the collector and earth of one of the transistors, say T_1, is observed by connection of a cathode ray oscillograph. On the waveform (Figure 4.31(b)) pq shows the rapid rise of C_{ce} from -8 V to zero, followed by a constant voltage represented by qr during the time that the transistor T_1 is saturated followed by the subsequent exponential decrease along rs as C_1 discharges through R_3. The cycle is then repeated. This waveform is approximately of rectangular shape.

It is also interesting to observe the variations of the base-emitter voltage V_{be} of one of the transistors. If a double-beam oscillograph is available, V_{ce} and V_{be} variations can be recorded together. Finally, the change in the period as R_3 and R_4 (maintained equal) are varied is noted.

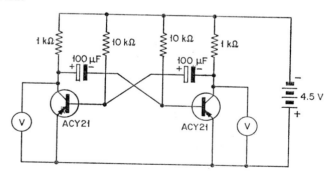

Figure 4.32. A multivibrator of long period (very low fundamental frequency)

A very long-period multivibrator (Mullard Ltd.) utilizing coupling capacitances $C_1 = C_2 = 100\ \mu\text{F}$ with $R_3 = R_4 = 10\ \text{k}\Omega$ enables the voltage variations V_{ce} to be observed with moving-coil voltmeters (Figure 4.32).

4.17 A Simple Transistor Tester

Figure 4.33 shows a simple unit which can be easily and cheaply constructed and which is most useful for checking quickly whether or not a low-power transistor or a diode is useful or useless. A 6 V battery supply is used. For a p-n-p transistor the polarity is as shown in Figure 4.33; for a n-p-n transistor this polarity must be reversed.

S_1 and S_2 are press-button switches which are closed when depressed. S_1 connects the battery supply; S_2 short circuits the resistor of 5.6 kΩ. A centre-zero milliammeter reading 1 mA − 0 − 1 mA is in series with the collector of the transistor together with a protective resistor of 330 Ω. R_s is a shunt resistor which, when switched-in by closing the switch S_3, alters the full-scale deflection of the milliammeter from 1 mA to 10 mA. S_3 is a double-pole, single-throw (DPST) switch; closing it also introduces the 150 kΩ resistor between the battery terminal and the base connection of the transistor.

If the transistor is assumed to be short circuited between its collector and emitter, with a 6 V supply connected (i.e. S_1 depressed and S_2 open) the total resistance in the circuit is (5.6 + 0.33) kΩ = 6 kΩ approximately. The maximum current that can be

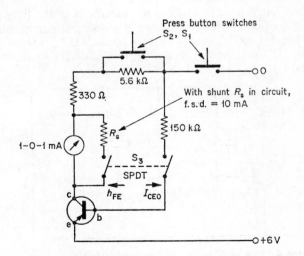

Figure 4.33. A simple transistor tester

passed with S_2 open is therefore $6A/(6 \times 10^3) = 1$ mA, and the meter cannot be damaged. The test sequence is as follows:

(a) With the switch S_3 in the I_{CEO} position, plug in the transistor and press switch S_1. The 150 kΩ and the shunt resistor R_s are both out of the circuit. There is no voltage on the base so the open-circuit collector-emitter current (I_{CEO}) is indicated.

If the deflection recorded by the meter is near full-scale (it can reach 1.0 mA) the transistor is useless. If the deflection is very small, press switch S_2. The 5.6 kΩ series resistor is then short-circuited so now I_{CEO} is measured. For a useful germanium transistor, I_{CEO} should be less than 600 μA; for a silicon transistor, I_{CEO} should be negligible.

(b) Switch S_3 across to the h_{FE} position. R_s and the 150 kΩ resistor are now both in the circuit so that the full-scale deflection at the meter is 10 mA and the base of the transistor is suitably connected.

With S_1 closed, the base current I_b is approximately $(6/150\,000)A = 40$ μA. Close switch S_2 (short-circuiting the resistor of 5.6 kΩ) and read the collector current I_c. Now,

$$h_{FE} = \frac{I_b}{I_c} = \frac{\text{meter reading in microampere}}{40 \text{ μA}}$$

As R_s has been chosen to give a full-scale deflection of 10 mA, the maximum value of h_{FE} which can be recorded is 10 mA/40 μA = 250.

(c) A diode can be tested by connecting it between terminals c and e, with S_3 switched to the I_{CEO} position. If the diode is satisfactory, when the switch S_1 is closed, the current flow will be negligible with the diode one way round and full-scale deflection will be recorded when the diode connections are reversed. A full-scale deflection recorded on closing S_1 whichever way round the diode is connected indicates that the diode is useless.

The resistor of 330 Ω in the circuit prevents the power that can be dissipated in the transistor from reaching values that could damage it. For example, for a transistor with a very large value of h_{FE}, say 250, the collector current would be 10 mA. The potential difference across the transistor is then

$$(6 - 330 \times 0.01) = 2.7 \text{ V}$$

and the corresponding power dissipated in the transistor is only $2.7 \times 0.01 = 0.027$ W = 27 mW, which is safe. A similar calculation will show that transistors with a lower value of h_{FE} will dissipate even less power.

Exercise 4

1. Explain, with appropriate diagrams, how you would investigate that static characteristics of a p-n-p junction transistor in (a) common-base; (b) common-emitter connection.
   Sketch the curves which are obtained and comment on their special features. (A.E.B.)
2. Define the hybrid parameters of a transistor in common-emitter connection. How would you determine their values for a p-n-p germanium transistor?
3. (i) Draw a sectional diagram to show the structure of a p-n-p germanium transistor in which the p-type regions have been produced using indium pellets. Comment on the relative size of the collector and emitter, and on the doping and the width of the n-type base region.
   (ii) Explain why a p-n-p junction transistor can be used as an amplifier in common-base connection when the collector current i_c is slightly less than the emitter current i_e.
   (iii) If the relationship between the collector current i_c and the emitter current i_e is given by $i_c = \alpha i_e$ and that between the collector current i_c and the base current i_b is given by $i_c = \beta i_b$, where α and β are constants, derive an expression relating α and β. ($\alpha = h_{FB}$, $\beta = h_{FE}$). (A.E.B.)
4. Derive an expression relating the leakage currents I_{CBO} and I_{CEO}.
5. Use a load-line to explain the terms *saturated* (i.e. *bottomed*) and *cut-off* for the a.c. operation of a p-n-p transistor in common-emitter connection.
6. Define the decibel and outline the advantages of using a logarithmic scale to express the power, voltage and current gain of an amplifier.
7. Assuming the load resistance in a common-emitter transistor amplifier to be small compared with the output resistance, draw a simple equivalent circuit for the amplifier. Explain how you could use this circuit to calculate the voltage gain, the current gain and the power gain of the amplifier.
8. Draw a circuit diagram of a common-emitter amplifier with resistive loading and a potential divider type of bias. Explain the action of the various components and the advantages of this type of bias.
9. Sketch the input, output and transfer characteritics of a p-n-p junction transistor in common-emitter connection.
   Draw the circuit diagram of a single-stage audio-frequency amplifier using a transistor in common-emitter connection. Comment on the problem of biasing the transistor. Outline features which differ from those of the equivalent valve amplifier. (A.E.B.)
10. Outline the methods of biasing a transistor and explain the advantages and disadvantages of each.
11. What is meant by (a) *positive feedback* and (b) *negative feedback*? In what type of circuit and for what purpose would each be used?

In a voltage amplifier of open-loop gain A a fraction of the output voltage is fed back to the input in antiphase with the input signal. Derive an expression for the gain of this feedback amplifier and use this expression to discuss the merits of negative feedback.

12. What special features would you associate with a constant current source?

    Draw a circuit diagram and explain the action of the components in a source capable of passing a constant current of about 5 mA through a thermistor of which the resistance varies between 10 Ω and 2000 Ω.

13. Explain the principles involved in providing a stabilized voltage supply by the use of
    (a) a Zener diode, and
    (b) a series control transistor.

14. Draw a circuit diagram and explain the working of a stabilized voltage supply of which the output voltage can be varied and which makes use of a difference amplifier.

15. Draw the circuit diagram and explain the operation of a transistor oscillator which provides a sinusoidal output. How would you measure the frequency of the oscillator?

16. Draw the circuit diagram and explain the principle of operation of a free-running multivibrator which utilises two p-n-p transistors.

17. With the aid of a suitable circuit diagram, explain the operation of a simple transistor tester which is capable of indicating leakage current and a value of h_{FE} of a p-n-p transistor.

18. A p-n-p junction transistor used as a voltage amplifier in common-emitter connection has hybrid parameters $h_{ie} = 1$ kΩ, $h_{oe} = 5 \times 10^{-5}$ siemen and $h_{fe} = 50$. The load resistance is 2 kΩ and the input e.m.f. is 20 mV. Draw an equivalent circuit for this amplifier and calculate the output voltage and output power.
[2.0 V; 2.0 mW].

5 Other semiconductor devices

5.1 The Unijunction Transistor

This type of transistor (UJT) consists of a bar of n-type silicon with ohmic base contacts B_1 and B_2 at the ends (Figure 5.1(a)). The resistance between these base contacts B_1 and B_2 (the *inter-base resistance*) is usually between 5 kΩ and 10 kΩ. This interbase resistance has a positive temperature coefficient which is approximately

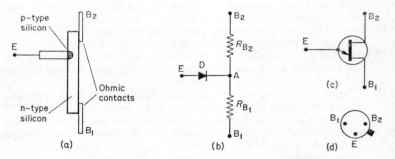

Figure 5.1. The unijunction transistor (UJT)

constant. This indicates that all the impurity atoms are ionized at room temperature and that the mobility of the carriers decreases with increase of temperature. An aluminium wire is joined by alloying to the silicon bar. As aluminium is trivalent, a p-n junction is thereby produced at the alloyed contact. This junction is usually nearer to the base B_2 than to base B_1. The emitter contact E is to this aluminium wire.

The equivalent circuit of the UJT (Figure 5.1(b)) consists therefore of the emitter terminal E joined by a p-n junction (diode) D to the junction A between two resistances R_{B1} and R_{B2} where $(R_{B1} + R_{B2})$ equals the interbase resistance. The circuit symbol and the base contacts are shown in Figure 5.1(c) and (d) respectively.

In normal use the *interbase voltage* V_{BB} is less than 30 V with B_2 positive with respect to B_1. The voltage at the point A (to which the emitter E is joined *via* D) with respect to B_1 is clearly given by

$$V_A = \left[\frac{R_{B1}}{R_{B1} + R_{B2}}\right] V_{BB} = \eta V_{BB}$$

where the constant fraction

$$\frac{R_{B1}}{R_{B1} + R_{B2}} = \eta$$

is called the *intrinsic stand-off ratio*.

If the emitter electrode E is connected to base B_1, the emitter-base diode (Figure 5.1(b)) is reverse biased with a voltage $\eta V_{BB} + V_D$

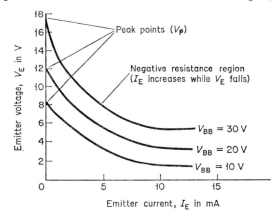

Figure 5.2. Emitter characteristics of a unijunction transistor

across the junction, where V_D is the equivalent diode voltage (see Figure 3.4). As the voltage V_E on the emitter is increased, current will not flow until the junction is forward biased, i.e. until $V_E = \eta V_{BB} + V_D$. The emitter voltage V_E at which this occurs is called the peak point value V_P and the equation is written

$$V_P = \eta V_{BB} + V_D \tag{5.1}$$

When $V_E = V_P$, holes are injected into the n-type silicon and, because B_2 is positive with respect of B_1, the holes are swept into the base B_1 region. Consequently, the resistance R_{B1} decreases rapidly. Likewise the voltage drop between the emitter and base B_1 falls, causing a further increase of the emitter current.

Characteristic curves for a UJT are drawn of the emitter voltage V_E against the emitter current I_E for a given interbase voltage V_{BB} (Figure 5.2). Once the emitter has reached the peak point voltage

V_P the emitter current I_E increases while the emitter voltage falls. The effective resistance of the device is then negative.

5.2 The UJT Relaxation Oscillator

All relaxation oscillators involve the charging of a capacitance C through a series resistance R, followed by a rapid discharge which returns the capacitor to its original state when the process is repeated.

A simple yet stable relaxation oscillator can be readily constructed with a unijunction transistor with a 9 V battery to provide the interbase voltage V_{BB} (Figure 5.3(a)).

The capacitor C is joined in series with a variable resistor R, of maximum value 50 kΩ to 100 kΩ. On switching in the 9 V supply,

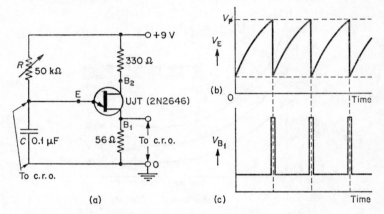

Figure 5.3. A unijunction relaxation oscillator

the capacitance C will charge up via the resistor R. When the voltage across C reaches the peak point voltage V_P, the emitter p-n junction becomes forward biased. The capacitor then discharges rapidly through base B_1 which is connected to the negative terminal of the battery (conveniently earthed) via a resistor of 56 Ω. With a UJT type 2N2646, this discharge of C continues until the voltage across it falls to approximately 2 V. At this point the emitter junction is no longer forward biased; no more holes are injected into the n-type silicon and the UJT switches off. The cycle is repeated as the capacitance C charges again via the resistance R.

Increase of R or increase of C will increase the period T of the oscillations, which is given approximately by

$$T = RC \log_e \left[\frac{1}{1-\eta} \right] \tag{5.2}$$

Other Semiconductor Devices

If the resistor R is decreased to below about 3 kΩ, the circuit ceases to oscillate and current flows continuously through the emitter-base junction. To obtain a long period T, large values of R and C are necessary: leakage currents through the capacitance and through the reverse-biased emitter junction can then cause unreliable behaviour.

Waveforms of voltage variations produced by the oscillator are readily displayed on a cathode ray oscillograph. Connection of the Y-plates of the oscillograph across the capacitance C shows the saw-tooth waveform due to the exponential charge of C followed by the rapid discharge (Figure 5.3(b)). Connection of the oscillograph across the 56 Ω resistor in series with base B_1 shows the sharp positive-going pulses brought about by the rapid discharges of the capacitance C (Figure 5.3c). A sharp negative-going pulse is shown by connecting the oscillograph across base B_2 and the negative terminal (usually earthed) of the battery.

This basic unijunction relaxation oscillator may be used as a timing circuit, a pulse generator or as a saw-tooth wave generator.

5.3 A UJT Staircase Generator

A second easily-constructed circuit based on a unijunction transistor demonstrates simply an electronic switch and the features of a frequency divider network (Figure 5.4).

The part of this circuit abcd is basically the same as that of

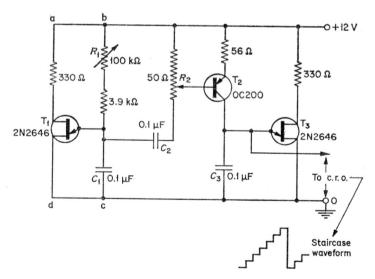

Figure 5.4. A UJT staircase generator

Figure 5.3, except for the values of the resistors used: it is a UJT relaxation oscillator. From the study of section 5.2 it is seen that negative-going pulses are generated across the resistance R_2 which is coupled to the relaxation oscillator *via* the 0.1 μF capacitance C_2 as shown. The repetition frequency of these pulses is determined by the magnitude of the resistance R_1 (which is variable). These pulses are fed to the base of the bipolar junction transistor T_2 of which the collector is connected to earth by the capacitance C_3 (0.1 μF). The amplitude of these pulses appearing at the base of transistor T_2 with respect to earth is decided by the setting of the tapping point on R_2. Initially, the emitter and base of the p-n-p transistor T_2 are at the same potential; arrival of a pulse from R_2 triggers T_2 into a conducting state.

The current pulses from the collector of transistor T_2 charge up C_3 in steps. Thus, each pulse creates a charge of Q on C_3 so that the voltage across C_3 increases to Q/C_3 with the first pulse, $2Q/C_3$ with the second, $3Q/C_3$ with the third, and so on. When the voltage across C_3 reaches the peak point voltage of the unijunction transistor T_3, this transistor is fired (rendered conducting) and C_3 is rapidly discharged.

The magnitude of R_1 determines the repetition frequency of the pulses and so the frequency of the steps; the magnitude of R_2 decides the pulse height and so the number of steps per cycle, i.e. the number required to charge C_3 up to the firing voltage of the unijunction transistor T_3.

Connection of the Y-plates of a suitably adjusted oscillograph across the emitter of the unijunction transistor T_3 and earth demonstrates the 'staircase' waveform. Together, the bipolar transistor T_2 and the unijunction transistor T_3 are seen to form a *frequency divider stage*. The number of steps per cycle (the number of steps in the staircase) may be adjusted from 10 to 100 or more. Similar stages may be added for further division, but the leakage of both the capacitors and the transistors will limit the number of stages able to function reliably.

Example 5.3

Values are given below of the peak point voltage V_p *for various values of the interbase voltage* V_{BB} *for a unijunction transistor. Plot a graph of* V_p *against* V_{BB} *and determine from it a value for* η, *the intrinsic stand-off ratio, and* V_D, *the equivalent diode voltage.*

V_{BB} (volt)	6.0	8.0	11.0	14.0	17.0	20.0	23.0
V_p (volt)	4.1	5.3	7.0	8.7	10.4	12.0	13.7

The values of V_p and V_{BB} listed in the table yield a straight line graph in accordance with equation (5.1). From the graph it is easily determined that $\eta = 0.57$ and $V_D = 0.7$ V.

5.4 The Silicon Controlled Rectifier (SCR)

As the name suggests, a silicon controlled rectifier is similar to a conventional diode but has a third control electrode called a *gate*. The circuit symbol (Figure 5.5) gives some idea of its behaviour. As with conventional diodes, it will block the flow of current when reverse biased. Of particular interest is the fact that it will not conduct even when forward biased unless triggered into a conducting state by a signal at the gate electrode.

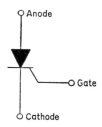

Figure 5.5. Circuit symbol for a silicon controlled rectifier

This firing signal, which must make the gate positive with respect to the cathode, may be a small d.c. current but more often is a current pulse because, once the device has been triggered into a conducting state, the gate no longer has any control.

The SCR can be switched off either by applying for an instant a reverse bias across the anode-cathode or by decreasing the current through the diode below a critical value called the *'holding current'*.

The SCR is one member of a family of four layer (p-n-p-n) devices called *thyristors*. The structure and physics of the SCR will not be described here. Instead, the concern is its special characteristics which enable a control signal of a few tens of milliwatts to switch power a million times as great. With a power gain of 10^6, the SCR is one of the most sensitive power control devices available.

5.5 A.C. Phase Control

The technical problem frequently arises that a load is supplied with electrical power from an a.c. supply and it is required to vary the amount of this power. The simple way of doing this is to connect the a.c. supply to the load *via* a variable series resistance (rheostat). For example, dimming of an electric lamp can be arranged by increasing the value of such a series resistance by simply sliding its tapping point. This method is wasteful because power is dissipated as heat in the series resistance.

In *phase control* this required variation of power supplied to a load is achieved by arranging to pass current through the load for only a fraction of each cycle of the a.c. supply. If, for example, the a.c. supply is chopped in some way so that only a quarter of each cycle of current passes through the load instead of the whole cycle, the power dissipated in the load is clearly much less. Furthermore, if this fraction is not simply a quarter but can be varied to be *any* fraction, power control is achieved.

The arrangement needed is a means of switching the a.c. supply on and off rapidly, and obviously more rapidly than the supply frequency. This switching to achieve phase control can be done admirably with a silicon controlled rectifier. A signal at the gate electrode at some chosen instant in the cycle will trigger the SCR into a conducting state and current will flow for the remainder of that cycle. The next cycle of the a.c. supply imposes a reverse bias on the SCR so that it will switch off the a.c. to the load.

Unlike the system which utilizes a rheostat for control, no power is dissipated in the control mechanism (the SCR). The SCR therefore serves as a very efficient means of controlling the power supply to lamps, heaters and electric motors.

The basic SCR phase-control circuit (Figure 5.6(a)) consists of an a.c. supply (often the mains supply at 240 V r.m.s. at 50 Hz) across which is the load resistance R_L (the device to be operated, e.g. lamp, heater or electric motor) in series with a silicon controlled rectifier. The control signal voltage is connected across the gate and cathode of this SCR.

The SCR can only conduct when its anode is positive with respect to its cathode. Assuming that the gate action is neglected, and the SCR is considered to be a simple rectifier, the current through R_L will be in the form of half-wave rectified pulses. Figure 5.6(b) represents one of the half-wave forms. This occupies half a cycle extending over 180°. At the beginning of this half-cycle, the phase angle α is conveniently considered to be zero, so it is 180° at the end.

In fact, even when its anode is positive, the SCR will only conduct if its gate electrode is made sufficiently positive with respect to its cathode. Suppose the gate voltage is thus triggered when $\alpha = 120°$. Conduction through the load R_L than takes place only when the phase angle is between 120° and 180°. Beyond 180°, the anode becomes negative in the succeeding half-cycle and conduction cannot occur again until the 120° position in the following positive-going half-cycle is attained. The angular duration of conduction is $180° - 120° = 60°$; this is called the *conduction angle*.

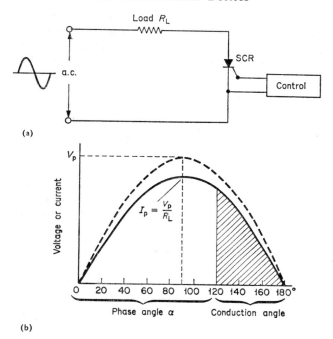

Figure 5.6. Principle of the SCR control circuit

If the load is purely resistive, the current is in phase with the applied voltage. The instantaneous current i at a time t is given by

$$i = I_\mathrm{P} \sin \omega t = \frac{V_\mathrm{P}}{R_\mathrm{L}} \sin \omega t$$

where $\omega = 2\pi f$, f is the supply frequency, V_p is the peak voltage across the load resistance R_L, and I_p is the corresponding peak current.

In a time $\mathrm{d}t$, a quantity of electricity $i\,\mathrm{d}t$ will flow through R_L. This current will only flow when ωt is between α and $180°$ (π radian). The period T of the a.c. corresponds to an angle of 2π radian. The current therefore only flows between the times $\alpha T/2\pi$ and $T/2$. During this interval, the quantity of electricity Q involved is given by

$$Q = \int_{\frac{\alpha T}{2\pi}}^{\frac{T}{2}} i\,\mathrm{d}t = \int_{\frac{\alpha T}{2\pi}}^{\frac{T}{2}} I_\mathrm{p} \sin \omega t\,\mathrm{d}t = \frac{I_\mathrm{p}}{\omega} \int_\alpha^\pi \sin \omega t\,\mathrm{d}(\omega t)$$

$$= \frac{-I_\mathrm{p}}{\omega} \Big[\cos \omega t\Big]_\alpha^\pi = \frac{-I_\mathrm{p}}{\omega}\Big[-1 - \cos \alpha\Big] = \frac{I_\mathrm{p} T}{2\pi}\Big[1 + \cos \alpha\Big]$$

This quantity of electricity Q is supplied during a fraction of the period T of one cycle; during the remainder of any particular period T, the quantity concerned is zero. The average current I_{dc} (i.e. that recorded by a moving-coil meter) which produces the charge Q in the time T is given by

$$I_{dc} T = Q = \frac{I_p T}{2\pi} (1 + \cos \alpha)$$

Hence $\qquad I_{dc} = \frac{I_p}{2\pi} (1 + \cos \alpha) = \frac{V_p}{2\pi R_L} (1 + \cos \alpha) \qquad (5.3)$

As the triggering action from the control device can alter the angle α, so I_{dc} can be varied.

By the use of a centre-tapped transformer in conjunction with two silicon controlled rectifiers, phase control of a full-wave rectified current through a load resistance R_L can be achieved (Figure 5.7).

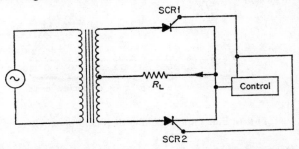

Figure 5.7. Control of full-wave rectified current through a load resistance by the use of two silicon controlled rectifiers

The control unit supplies pulses simultaneously to the gates of each SCR. Whichever SCR is forward biased when the firing pulse arrives will be triggered into conduction.

Two silicon controlled rectifiers with separate control units can be used to provide a load with full-wave a.c. phase control, i.e. phase control during each half-cycle of the supply (Figure 5.8).

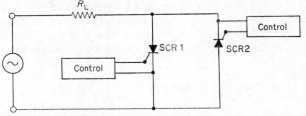

Figure 5.8. Full-wave a.c. phase control by the use of two silicon controlled rectifiers

5.6 A Bistable SCR Circuit

The term 'bistable' implies that a circuit can be put into either one of two stable states. The idea is illustrated simply by the use of an ordinary two-way switch in a circuit such that the battery of e.m.f. E can be connected either across the resistance R_1 or across the resistance R_2 (Figure 5.9). If the conducting arm of the switch S connects terminal 0 to terminal a, current flows through R_1 but not through R_2. This is one stable state. When the conducting arm is flipped over so that 0 is joined instead to b, current flows through R_2 but not through R_1. This gives the second stable state.

Transference from one stable state to another is achieved in this case by the mechanical action of 'throwing' the switch. This clearly

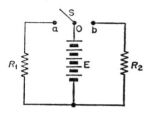

Figure 5.9. The concept of a bistable circuit

takes a finite time, say 0.1 second. In a bistable electronic circuit, the object is to achieve much more rapid switching from the one stable state to the other, e.g. in a microsecond or even a nanosecond in some cases. Furthermore, the 'throw' of the 'switch' is achieved at the behest of a pulse generator.

Thermionic vacuum tubes, transistors and other semiconductor devices have all been used in bistable circuits for various purposes (e.g. counting pulses, digital computers). A simple one requiring

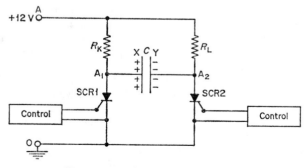

Figure 5.10. A bistable SCR circuit

few components and able to control large currents utilizes two silicon controlled rectifiers (Figure 5.10).

In this circuit assume that when the d.c. supply voltage is applied, neither SCR conducts initially although each is forward biased. When a firing pulse is applied by its control unit (a pulse generator) to the gate of SCR2, current will flow through it and through its load resistance R_L.

The voltage at point A in this circuit relative to earth is fixed by the battery supply at $+12$ V, (assume that the negative terminal of the supply is earthed for convenience: it often is, but does not have to be). There is no current through SCR1 and load resistance R_K so that point A_1 (the anode of SCR1) is also at $+12$ V. The current established through R_L causes a voltage drop across it: the voltage at point A_2 (the anode of SCR2) therefore falls to below 12 V. The capacitance C therefore becomes charged with its plate X at $+12$ V and plate Y at some positive voltage less than $+12$ V. The polarity of the charge on C is thus as shown in Figure 5.10.

A firing pulse now applied to the gate of SCR1 renders it conducting. Consequently, the charged capacitance C is connected across SCR2. This capacitor voltage establishes for an instant a reverse bias across SCR2 which stops it conducting (i.e. switches it off).

The circuit is now in a symmetrically reversed state from that existing initially in that SCR1 is conducting and SCR2 is not. Consequently, C becomes charged up with a reverse polarity from that which it had initially. The voltage across C is thus available to switch off SCR1 when a second pulse fires SCR2.

The circuit may thus be set in one of two stable states: with SCR2 conducting and SCR1 not conducting or *vice versa*. The transfer from one stable state to the other (i.e. the conducting roles of SCR2 and SCR1 reversed) is achieved by a pulse of a few milliwatt applied to the gate electrodes. The term bistable circuit is clearly appropriate.

The power delivered to the load resistance R_L is determined by the current through SCR2. This current can be large by choosing a heavy duty type for SCR2. The other silicon controlled rectifier SCR1 merely serves to provide the switching action and can hence be a smaller, cheaper component.

This bistable circuit hence enables large amounts of power through the load R_L to be switched on and off at the behest of a very small power signal applied to the gates of SCR2 and SCR1. A minimum value of the capacitance C is required for effective switching.

On occasions the resistance R_K is chosen to limit the current through SCR1 to a value below its holding current. Although SCR1 will switch on and serve to switch off SCR2, it cannot remain on and reverts almost immediately to a non-conducting state.

5.7 A Simple Demonstration of Phase Control

To arrange a simple demonstration of phase control with inexpensive apparatus, the circuit of Figure 5.11(a) may be easily constructed to trigger a SCR into conduction from an a.c. supply. A variable resistance R, a diode D_1 and a protective resistance R_G

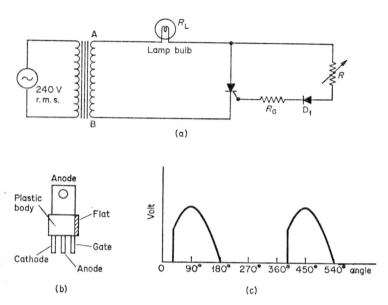

Figure 5.11. (a) A simple phase-control using an SCR (Suitable component values:— Transformer secondary: 12 V r.m.s., 1 A. Lamp bulb: 12 V, 0.75 W. R: 5 kΩ. Potentiometer. R_G: 50 Ω. SCR is type IRC10. D_1 is type 10D6) (b) the IRC10, and (c) the voltage waveform across the load R_L

in series are connected across the anode and gate electrode of this SCR. A load resistance R_L (conveniently a lamp bulb for demonstration) is in series with the SCR across the transformer secondary.

During the half-cycle when end A of the transformer secondary winding is positive with respect to the other end B, the SCR is forward biased. This positive half-cycle also supplies a current to the gate of the SCR. The magnitude of the current in this gate circuit will depend on the value of the resistance R. Variation of R hence enables the phase angle at which the SCR will switch on to be controlled between zero (full on) and 90°.

The protective resistance R_G is required to prevent the flow of excessive gate current when R is set at a minimum value. The diode D_1 is necessary to block the reverse gate voltage which would otherwise be obtained during the negative half-cycles of the supply voltage (i.e. A negative with respect to B).

When the gate current is sufficient for the SCR to conduct, a large current pulse passes through R_L. As R_L is much greater than the resistance of the conducting SCR, almost the whole voltage is across R_L. Consequently, the gate current drops to near zero. An oscilloscope connected across the load R_L shows the voltage wave form (Figure 5.11(c)).

In the circuit of Figure 5.11(a), a transformer has been used to isolate the circuit from the a.c. mains supply and so make it simple and safe to connect the oscilloscope across R_L.

5.8 An Experiment to Construct a Lamp Dimming Circuit

The components and component values required are noted on the circuit diagram (Figure 5.12(a)). The 6 V lamp bulb (R_L) is fed with full-wave rectified current from a step-down transformer T_1 which has a centre-tapped secondary winding providing $6 - 0 - 6$ V. The rectifiers needed are both silicon controlled types (ICR 10) so that the magnitude of the rectified current through R_L is controllable by pulses fed to their gate electrodes. This current control clearly enables the lamp to be dimmed.

The required pulses to the gates of the silicon controlled rectifiers are provided by a unijunction transistor relaxation oscillator (part of the circuit within abcd on Figure 5.12(a), *cf.* Figure 5.3). The voltage supply to this unijunction transistor circuit is from the power pack (within the dotted lines on Figure 5.12(a)) comprising the transformer T_2, diodes D_1 and D_2 (both of type IDP 10) and a Zener diode MZ22 for voltage limiting (section 3.17). Note that this power pack does not have a smoothing capacitor. The output pulses across the 56 Ω resistor of the relaxation oscillator between point A and earth are fed to the gate electrodes of the two silicon controlled rectifiers.

This circuit provides an interesting example of the need for *synchronization*. To provide effective phase control, the firing pulses to the silicon controlled rectifiers must be synchronized with the a.c. mains voltage. This is achieved because the circuit is fed with a full-wave rectified supply (see waveform at 1 on Figure 5.12(a)) of which the peak values are clipped by the Zener diode (see waveform 2 on Figure 5.12(a)). As, however, this supply voltage is *not* smoothed, the interbase voltage V_{BB} across the unijunction transistor (2N 2646) falls to zero at the end of each half-cycle. At this time, the emitter of the unijunction transistor conducts to discharge the capacitance C (0.1 μF). Hence the voltage across this capacitance is set to the same reference value at the end of each half-cycle. The

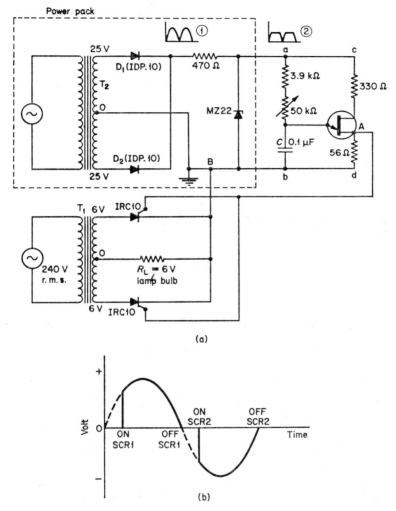

Figure 5.12. (a) A lamp dimming circuit and (b) the voltage waveform across the lamp R_L

timing circuit (i.e. the relaxation oscillator) then determines after what interval there occurs the output pulse which will fire the forward biased SCR The variable 50 kΩ resistor then allows the conduction cycle of the two SCRs to be varied between angles of 0° and 170° approximately.

The waveforms depicted at points 1 and 2 can be observed by connection of an oscillograph to either one of these points and earth.

Example 5.8

A silicon controlled rectifier with a load resistor of 100 Ω in series is fed with an a.c. supply of 100 V r.m.s. at 50 Hz. It is fired, by pulses to its gate electrode, at an angle of 60°. If the voltage drop across the conducting SCR is taken to be zero:

    (a) *calculate the average current that flows through the load resistor*;
    (b) *draw the voltage waveform across the load resistor*;
    (c) *draw the voltage waveform across the SCR*.

(a) The average current I_{dc} is given by equation (5.3):

$$I_{dc} = \frac{V_p}{2\pi R_L}(1 + \cos \alpha)$$

where V_p is the peak applied voltage, R_L is the load resistance and α is the phase angle. Hence

$$I_{dc} = \frac{100 \times 1.4}{2\pi \times 100}(1 + \cos 60°)$$

$$= \frac{1.4}{2\pi}(1 + 0.5) = 0.33 \text{ A}.$$

    (b) The voltage waveform across the load resistor is given in Figure 5.13(a).
    (c) The voltage waveform across the SCR is given in Figure 5.13(b).

5.9 Field Effect Transistors

The transistor described in Chapter 4 are called *bipolar* transistors because their performance depends on the interactions of two types of charge carrier: electrons and positive holes. The field effect transistor (FET) is a *unipolar* device because only one type of charge carrier is involved: electrons in an n-channel FET and holes in a p-channel FET.

The motion and concentration of these carriers within the semiconductor material of an FET are controlled by an electric field set up within the conducting region by voltages applied to contacts to the semiconductor material.

Here the thinking of the scientist might be said to be guided by previous experience with the triode valve. In a triode valve the electron current from the thermionic cathode is controlled by a grid which is normally negative in potential with respect to the cathode. This grid therefore does not attract electrons, so the current to it

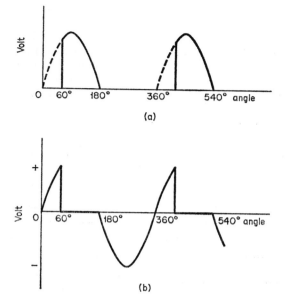

Figure 5.13. Voltage waveform across (a) the load resistor and (b) the SCR in Example 5.8

(the grid current) is negligibly small. By its electric field influence on the electron space charge it does, however, control the anode current as its negative voltage is varied. The input resistance to this control grid (which requires virtually zero current) is therefore very high. The input resistance to the bipolar transistor is low so significant current drain on the input device is demanded.

The problem is: can a semiconductor device be made with a very high input resistance? The solution is to make such a device in which the current carrier motion is controlled by electric field changes. As this has been achieved in the FET, yet another branch of electronics (that of high input impedance amplifiers) in which the thermionic vacuum tube had been supreme for some four decades has become dominated by semiconductor devices.

Early attempts to produce an electric field to control the current within a semiconductor by maintaining a voltage across insulated electrodes at or near the surface of the semiconductor were unsuccessful. In all cases, the surfaces appeared to screen electrically the interior of the semiconductor. Indeed, it was while investigating these surface effects that Shockley and his associates Bardeen and Brattain discovered transistor action in 1948 and eventually introduced the bipolar transistor, but with its low input resistance.

Some years later Shockley was responsible for introducing a solid state device with an input resistance of several hundred megohm. He realized that a controlling electric field could be created within the body of a semiconductor by the use of a reverse biased p-n junction. The device—often called a junction gate field effect transistor—depends for its action on the creation and control of a depletion layer produced at a reverse biased junction.

5.10 The Structure and Behaviour of an FET

The creation of a depletion layer has already been mentioned (section 3.3). Figure 5.14 shows the form of a depletion region for a p-n junction in which the semiconductor material in the p-type

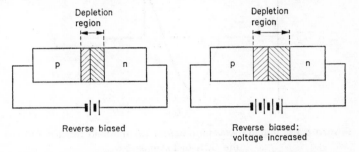

Figure 5.14. Effect of reverse bias on a depletion region about a p-n junction where the p-type region is more heavily doped

region has a much greater carrier concentration (is more heavily doped) than the material in the n-type region.

As the reverse bias is increased, the width of the depletion region increases, but it always extends further into the n-type material than into the p-type. This is because each type of material must contribute the same number of current carriers so that a greater volume of the less heavily doped n-type material is depleted. On the n-type side of this junction, the loss of electrons which have diffused across the junction leaves positive ions; on the p-type side, loss of holes which have diffused across the junction leaves negative ions. The ions are locked in the crystal lattice: they are not mobile. The conductivity of this depletion region is nominally zero because mobile current carriers are not available.

A simple model of an n-channel FET may be envisaged by considering a bar of n-type silicon with ohmic contacts at each end (Figure 5.15). If this bar has length l, width w and thickness t, the resistance R between these end contacts is given by

$$R = \rho l / wt$$

where ρ is the resistivity of the n-type silicon. A battery providing an e.m.f. of, say, 6 V is connected across the ends of this bar where the ohmic contact electrode at the negative end (usually earthed) is called the *source* S and the contact at the positive end is the *drain* D. The voltage across D and S, V_{DS}, is called the drain-to-source voltage.

Within each side of this bar of thickness t are produced layers of heavily doped p-type silicon. The electrodes GG attached to these layers are called the *gate*. Normally this gate is biased negatively

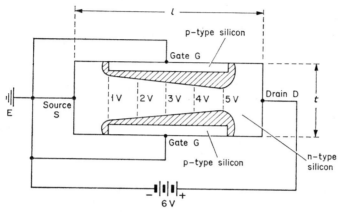

Figure 5.15. Simple model of the action of an n-channel FET

with respect to the source S, the gate-to-source voltage being V_{GS}, where V_{GS} is negative.

The current to the drain electrode, I_D, is due to the movement of electrons along the n-type silicon bar. The current to the gate electrodes, I_G, is vanishingly small if V_{GS} is negative because no mobile current carriers are available to cross the reverse-biased p-n junctions. The drain current I_D is a function of both the drain-source voltage V_{DS} and the gate-source voltage V_{GS}.

The important characteristics of the FET are hence the drain current I_D plotted against the drain-source voltage V_{DS} for various fixed values of the gate-source voltage V_{GS}. A given value of V_{GS} which is negative will produce a reverse bias across the two p-n junctions on either side of the silicon bar of which the widths of the depletion regions depend on the magnitude of V_{GS}. An increase from zero of the drain-source voltage V_{DS} (with V_{GS} constant) will have initially a two-fold effect:

(*a*) it will cause the drain current I_D to increase on the presumption that the n-type silicon bar is simply an ohmic conductor so the velocities of the electrons will increase;

(b) it will also increase the width of the depletion regions about the p-n junctions associated with the gate electrodes: this will cause I_D to decrease.

The statement (b) needs to be examined further. Suppose that V_{DS} is 6 V. Ignoring for the moment the gate-source voltage V_{GS}, it is clear that there will be a distribution of voltage along the length of the silicon bar due to V_{DS}. This distribution will increase from zero at the source to 1, 2, 3 etc. up to 6 V at the drain (Figure 5.15). The voltages all act as reverse bias values adding to the gate-source voltage, making it effectively more negative. The maximum negative bias, and so the maximum depletion region width, will exist over a small region near the drain-end of the gate electrodes.

This increase of the depletion region widths reduces the effective conducting cross-section of the n-type silicon bar. A *pinch-off effect* is obtained: the electrons cannot travel so readily to the drain through this narrow channel, so the drain current I_D decreases.

With a given value of V_{GS}, the effect of increasing the drain-source voltage V_{DS} is a consequence of both actions (a) and (b). The result is that the drain current I_D increases at first almost linearly as V_{DS} is increased from zero. When V_{DS} is a few volts positive, however, the effect (b) becomes more and more pronounced, resulting in the fact that I_D becomes independent of a further increase of V_{DS}. Thus, beyond a certain value of V_{DS}, I_D is constant.

It is not easy to prove that I_D becomes a constant independent of V_{DS}. This can be shown to be the case by a more intimate study of the voltage gradient in the vicinity of the narrowest part of the channel between the depletion regions, but involves a study of the mobility of the electrons which becomes a function of the electric field strength (voltage gradient) at high values and will not be attempted here.

The effect on the drain current I_D of increasing the negative bias on the gate, V_{GS}, for a given value of the drain-source voltage V_{DS} is that it increases the width of the depletion regions around the p-n junctions. This causes I_D to decrease. If V_{GS} is made sufficiently negative, the pinch-off effect of these depletion region widths is great enough nominally to extend right across the silicon bar and reduce I_D to zero.

The terms which have been used to relate specifically to the FET are:

(i) *The source S*: the electrode by which the majority carriers enter the bar.
(ii) *The drain D*: the electrode by which the majority carriers leave the bar.

(iii) *The gate G*: the heavily doped p-region (often denoted by p⁺, see section 2.2) is called the gate electrode because of its controlling function.
(iv) *The channel*: the region in the bar between the two gate electrodes. The majority carriers move through this channel from the source to the drain. Electrons are the majority carriers in an n-channel FET (one based upon an n-type silicon bar), whereas positive holes are the majority carriers in a p-channel FET.

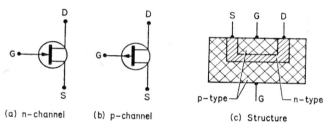

(a) n-channel   (b) p-channel   (c) Structure

Figure 5.16. Circuit symbols and structure of an FET

Because the mobility of holes in silicon is only half that of electrons, the p-channel FET has inferior characteristics to an n-channel one of the same geometry. Consequently, n-channel FETs are the much more widely used.

In Figure 5.16 are shown the circuit symbols for an n-channel FET,

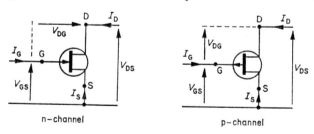

n-channel   p-channel

Figure 5.17. Polarity conventions in relation to the operation of n-channel and p-channel FET's

a p-channel FET and a practical FET structure. Certain polarity conventions are illustrated by Figure 5.17. If the polarity of the current or voltage is opposite to that indicated by the arrow, the value noted must carry a negative sign. Hence an n-channel FET operating in a normal manner might have a drain-source voltage V_{DS} of $+10$ V, and a drain current I_D of $+1$ mA with a gate-source bias V_{GS} of -2 V. On the other hand, a p-channel FET operated under similar conditions would have $V_{DS} = -10$ V, $I_D = -1$ mA and $V_{GS} = +2$ V.

5.11 An Experiment to Plot the Drain Characteristics of an n-channel FET

In the circuit employed (Figure 5.18(a)) the drain-source voltage V_{DS} can be set and recorded at values between 0 and 20 V, the drain

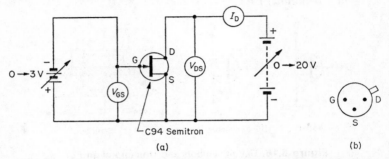

Figure 5.18. Determination of the drain characteristics of an n-channel FET

current I_D is recorded by a milliammeter (0 — 10 mA), and the gate-source voltage V_{GS} can be set and recorded at 0, -1, -2 and -3 V.

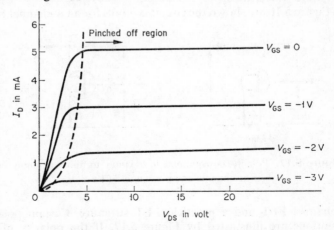

Figure 5.19. FET drain characteristics

An n-channel FET of type Semitron C94 is used. The base connections to the FET are shown in Figure 5.18(b).

The drain characteristics (I_D against V_{DS} for various values of the gate-source voltage) shown in Figure 5.19 exhibit two distinct regions of particular interest:

(i) *Below pinch-off*: in this region the drain current I_D increases with the drain-source voltage V_{DS}. The device acts like a variable resistance of which the magnitude is controlled by the gate-source voltage V_{GS}. Some types of FET are often used as voltage-controlled resistors.

(ii) *The pinched-off region*: the drain current I_D is virtually independent of the drain-source voltage V_{DS}. It is in this region that the FET is normally operated when used as an amplifier.

The *transfer characteristic* (I_D against V_{GS} for a given value of V_{DS}) can also be plotted from the results obtained from the circuit of Figure 5.18. A typical transfer characteristic for an n-channel FET is shown in Figure 5.20.

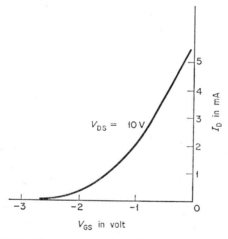

Figure 5.20. The transfer characteristic of an n-channel FET

5.12 A Simple FET Voltmeter

In many experiments on transistors and also in several other electrical measurements it is essential to have available a millivoltmeter with a very high input resistance, preferably exceeding 10 MΩ. For most student experiments, the high-quality but rather expensive instruments available from manufacturers are not generally required. The circuit diagram of an inexpensive FET voltmeter* which is ideal for

---

* The design and constructional details of this instrument are given by the *Mullard Educational Service* in their series of booklets: 'Educational Projects in Electronics'.

general use and easily constructed in a school or college laboratory is shown in Figure 5.21.

This instrument makes use of the inherently high gate-source resistance of an n-channel filed effect transistor (type BFW16). The input resistance is further increased and a linear response is achieved by providing negative feedback *via* the resistance R_s in the source lead. The FET may now be regarded as a voltage operated device demanding negligible current from the input circuit. The effective resistance between the gate G and the zero voltage (earth) line is so

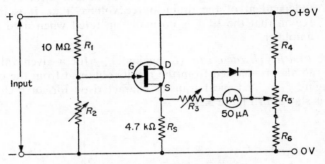

Figure 5.21. A simple FET voltmeter

high that its shunting effect on the resistor R_2 is negligible, even when R_2 is as high as 40 MΩ.

With the input terminals shorted, i.e. $V_{in} = 0$, a current (I_{DS}) of, say, 0.50 mA flows through the channel, i.e. between the drain and the source. Due to the resistance R_s of value 4.7 kΩ, the source is at a positive potential of $(0.5 \times 10^{-3} \times 4.7 \times 10^3) = 2.35$ V with respect to the zero voltage line. This steady current (I_{DS}) conveniently establishes the bias condition of the FET, namely that the gate electrode is -2.35 V with respect to the source (the gate-channel junction is reverse biased). The tapping point on R_5 (with $V_{in} = 0$) is adjusted to give a zero reading on the microammeter. With a positive input of 200 mV applied on the gate relative to the zero voltage line, the resistance R_3 is adjusted to give a full-scale deflection (50 μA) on the microammeter. Once R_5 and R_3 (relating to the zero and full-scale deflection of the microammeter), have been set they are not altered again.

The steady input of 200 mV required can be obtained from a low resistance source, for example, a potential divider across a 2 V cell and measured with a calibrated potentiometer or a digital voltmeter.

Resistors R_1 and R_2 form a potential divider across the input to this FET voltmeter. As the meter is adjusted to give a full-scale deflection for an input of 200 mV, the value of R_2 may be chosen by means of a range switch to give a full-scale deflection of, say, 1.0 V, 10.0 V and 100 V. The value of R_2 required in each case is readily calculated.

5.13 Special Features of FETs

(a) The operation depends on the motion of majority carriers (electrons for an n-channel FET) only. This unipolar behaviour is less noisy (i.e. less liable to small random current fluctuations) than that of the bipolar transistor.

(b) The input resistance is high: typically several megohm.

(c) They are relatively immune from nuclear radiations and consequently are often used in preamplifiers in radiation detectors.

(d) For an n-channel FET, the channel current is decreased when the negative voltage applied to the gate with respect to the source is increased, and it is fully conducting when the gate-source voltage is zero.

5.14 Semiconductor Particle-Detectors

In the experimental study of radioactivity and nuclear physics generally much use is made of devices able to detect alpha-particles, beta-particles and gamma-rays. The α-particles are helium nuclei (doubly positively charged helium ions) with speeds usually up to about $0.1c$ (c being the speed of light in free space), β-particles are electrons with speeds up to $0.9c$ and more, whilst γ-rays are photons of energy $h\nu$ $(=hc/\lambda)$ where the wavelength λ is less than about 0.01 nm and h is the Planck constant (6.625×10^{-34} joule second).

The particles are detected by ionization chambers, gas-filled counter tubes such as proportional counters and Geiger–Müller counters, fluorescent screens leading to the scintillation counter with a photomultiplier tube, photographic emulsions, cloud chambers, bubble chambers and the like. Some of these devices, particularly with the aid of electronic circuit methods, also lend themselves to the determination of the energies of the particles and their energy spectrum, i.e. out of the total number of particles N emitted per second, the number of particles ΔN having a particular energy between E and $(E + \Delta E)$ is found.

Semiconductor detectors are gradually replacing these more conventional devices. Their response is more rapid than that of the gas-filled ionization devices and even than that of the fluorescent devices, so they can count greater numbers of particles per second, they are able to distinguish between identical particles close together in energy and they are often simpler and more convenient to use.

The ability of a device to distinguish between two particles close together in energy (so providing high *energy resolution*) is determined by the energy required to produce the phenomenon which results in an output signal from the device.

In this connection it is interesting to compare the creation of an electron-hole pair in silicon or germanium with the creation of a pair of ions (a positive ion and an electron) in a gas. Whereas the former requires an energy of about 3 eV, the latter demands about 30 eV. Furthermore, in the solid semiconductor, the energy of the incident particle to be detected is absorbed in a short path because the atoms are only a few tens of nanometres apart whereas in a gas at atmospheric pressure, the atoms (or molecules) are separated by mean distances of a few hundreds of nanometres, and this mean distance increases inversely with the gas pressure. In a gas in an ionization chamber (or gas-filled counter tube) a considerable path length of the incident particle is therefore needed to produce an adequate ionization current or current pulse. Consequently, gas-filled particle detectors are considerably larger in size than solid-state or semiconductor particle detectors.

Although a variety of semiconductor particle detectors exist, only the silicon surface junction type is considered here. The surface of a crystal of p-type silicon (the silicon is usually doped with trivalent boron as the acceptor impurity to make it p-type) is doped by diffusion with pentavalent phosphorus to create an n-type layer. Across the p-n junction thus created very near to the surface there exists a high resistance depletion region with an electric field directed from the n-type material (containing excess positive ions locked in the crystal lattice) to the p-type material (containing excess negative ions). The width of this depletion region is increased on the application across the p-n junction of a reverse bias voltage (section 3.3).

The particles to be detected are incident upon this p-n junction. If these particles are ionizing, on passage through the p-n junction they will create a trail of free electrons and positive holes in the depletion region. The electric field across the junction accelerates these free current carriers which may be collected in a few nanoseconds by suitably placed electrodes. The current pulse produced is a record of the passage of a particle.

The energy required to create an electron-hole pair in silicon is about 3.5 eV. In germanium, with its narrower forbidden energy gap (section 1.18), the required energy is even lower. However, germanium suffers from the disadvantage that it has a high noise level (produces random currents due to thermal effects) unless it is cooled to well below room temperature whereas silicon operates satisfactorily without cooling.

A difficulty with such semiconductor particle detectors is that the incident particle energy must be less than about 30 MeV, otherwise the depletion region is traversed so rapidly that little effect is created within it. Attempts are being made to increase the width of the depletion region so that considerably more energetic particles will give up all their energy on transit.

The structure and form of a typical commercial surface junction silicon particle detector is shown in Figure 5.22.

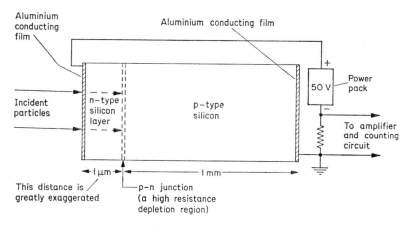

Figure 5.22. A surface junction silicon particle detector

5.15 Photoelectric Effects in Semiconductors

If photons of energy $h\nu$ are incident on the depletion region of a surface type p-n junction similar to that used in particle detectors (section 5.14), electron-hole pairs are created. The electric field existing across the depletion region causes motion of the free charges created. Equilibrium may be restored by current flow in an external circuit.

The energy gap for silicon is 1.1 eV (section 1.18). Photons with energies exceeding this value are therefore effective. The corresponding threshold wavelength λ_t of the incident radiation is hence given by

$$h\nu = hc/\lambda_t = 1.1 \text{ eV}$$

In this relationship h must be in eV second instead of the usual joule second. As $h = 6.625 \times 10^{-34}$ joule second and $1 \text{ eV} = 1.6 \times 10^{-19}$ joule, so

$$h = \frac{6.625 \times 10^{-34}}{1.6 \times 10^{-19}} = 4.13 \times 10^{-15} \text{ eV s}$$

Putting $c = 3 \times 10^8$ m/s,

$$\lambda_t = \frac{4.13 \times 10^{-15} \times 3 \times 10^8}{1.1} = 1.13 \times 10^{-6} \text{m} = 1130 \text{ nm}$$

The visible region of the spectrum extends in wavelength from 7×10^{-7}m (700 nm) to 4×10^{-7} m (400 nm) approximately. The semiconductor detector therefore gives a photoelectric effect with visible light and infra-red radiation up to wavelengths of 1100 nm approximately.

The photoelectric effect in this case is more specifically a *photovoltaic effect* in that the incident radiation on the p-n junction produces a change of the voltage across it.

A number of factors at present limit the efficiency of these photovoltaic semiconductor cells to about 10 per cent, i.e. the electrical power output is about 10 per cent of the energy per second (power) due to the incident radiation. These factors include:

(*a*) The electron-hole pairs created recombine before they are collected by attached electrodes.
(*b*) The radiation (light) is partly reflected from the front surface of the cell, and so some of it does not traverse the depletion region about the p-n junction.
(*c*) the electrical resistance of the cell causes power loss because passage of current through it results in dissipation by heat.

Nevertheless, surface type p-n junction silicon cells have become important not only as photovoltaic detector devices for radiation but also as *solar batteries*.

In full sunlight, such a silicon cell will develop an e.m.f. of approximately 0.6 V on open circuit (i.e. no load resistance present). The larger spacecraft and satellites frequently contain a solar cell panel containing about 5000 such cells to provide power of 100 W at 30 V. These cells are connected in rows in parallel to reduce resistance and with these rows connected in series to provide the output voltage.

Among the other devices which exploit the creation of electron-hole pairs by incident photons are *photodiodes* and *phototransistors*. The structure and approximate dimensions of a photodiode are shown in Figure 5.23(a). The diode is always operated with reverse bias, with up to 30 V across the p-n junction. The dark resistance (resistance with no incident light) at this maximum voltage is about 2 MΩ. The corresponding dark current is then 30 A/(2 × 10^6) = 15 μA. This current is obviously due to minority carriers.

A converging lens is normally used to focus the incident light at the photodiode in the otherwise light-tight capsule. This light therefore impinges in the immediate area of the p-n junction behind

Other Semiconductor Devices   167

the lens window. Alternatively, a separate, external lens and a plane window may be used. Electron-hole pairs are therefore created near the surface to form the reverse current which is now large compared with the dark current. This reverse current is a linear function of the illumination produced by the incident light. It might be limited to say, 3.0 mA to avoid temperature rise due to heating effects at the junction. The device can operate up to a frequency of 50 kHz, and thus can easily detect the fluctuations in light intensity from discharge tube lighting.

Operating in the manner described, the photodiode behaves as a current generator with a high internal resistance. A resistance in the external operating circuit has little effect on the current flowing (Figure 5.23(b)). The voltage developed across this external circuit has a magnitude dependent upon the illumination by the incident

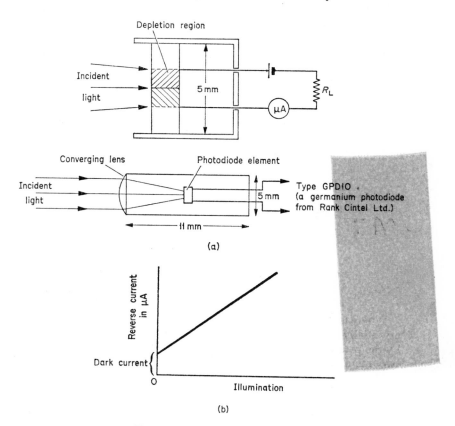

Figure 5.23. The photodiode

light. This voltage can be used to operate a circuit or device in light-controlled apparatus. An oscillograph connected across this resistance will display the waveform of any fluctuations in the source of light that produces the illumination of the photodiode.

The germanium photodiode type GPD10 (marketed by Rank Cintel Ltd.) referred to in Figure 5.23 has a peak response to radiation of wavelength 1500 nm in the infra-red but still responds well in the visible region.

A small filament lamp operated from a constant voltage transformer or a constant current source is ideal for experiments with photodiodes.

The *phototransistor* is simply a junction transistor in the capsule of which is a region through which light may be passed to impinge on the sensitive base region. The current carriers produced on incidence of light on the base region are amplified by the transistor action. The electron-hole pairs produced by the light are equivalent to base current; the number produced therefore controls the much larger current which flows between the emitter and collector.

The circuit of a linear light meter (Figure 5.24) described by Mullard Ltd. utilizes a BPX25 silicon planar phototransistor.

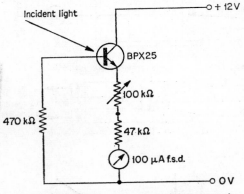

Figure 5.24. A phototransistor light meter circuit

Whereas solar batteries, photodiodes and phototransistors are essentially photovoltaic devices, a photoconductive cell is formed by the production of a thin film of semiconducting material between two metal electrodes (Figure 5.25). The secmiconductor material used in such photoconductive cells is cadmium sulphide (CdS) or lead sulphide (PbS) or indium antimonide (InSb) in the form of a thin polycrystalline film. Incident light traverses the transparent substrate and the photons impart energy to the semiconductor film and create in it electron-hole pairs which increase its conductivity.

The circuit used simply involves the photoconductive cell in series with a microammeter or milliammeter and a steady voltage supply. As the cell resistance changes depending on the light illumination, the current recorded by the series meter changes. For visible light, cadmium sulphide is mainly used. For infrared radiation, lead sulphide cells are effective.

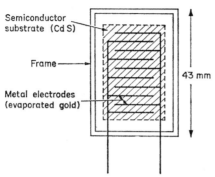

Figure 5.25. A photoconductive cell

The dark resistance of the CDS1 photoconductive cell is approximately 100 MΩ. The maximum voltage that can be applied across it is 300 V. The current through the cell is limited by a series resistor to prevent undue power from being dissipated in the cell. For the CDS1, the maximum power is 0.5 W.

When the illumination is removed, the current decays exponentially over a period of about one minute.

A widely used application of cadmium sulphide photoconductive cells is in exposure meters used in photography. The operating e.m.f. is a few volt. They are sufficiently sensitive to be able to record the exposure required even when the scene to be photographed is illuminated only by bright moonlight.

5.16 Thermistors

Thermistors are semiconducting resistors with a large negative temperature coefficient of resistance. Thus, as the temperature is increased, the resistance decreases. The temperature change can be caused either in the surroundings in which the thermistor is immersed or by heat generated within its element due to the passage of current through it.

Over a specified temperature range, the temperature coefficient of resistance, α, is defined as

$$\alpha = \frac{\text{rate of change of resistance with temperature}}{\text{original resistance}}$$

which, in mathematical terms, becomes

$$\alpha = \frac{1}{R_T} \frac{dR_T}{dT} \qquad (5.4)$$

where R_T is the resistance at a temperature T K. For thermistors at room temperature (20°C = 293 K), α has a value of -0.06 per degree Kelvin (-0.06 K^{-1}) which is -6 per cent K^{-1}. For metals, α is positive and about $+0.003$ K^{-1}, i.e. 0.3 per cent K^{-1}. The thermistor is clearly, therefore, much more sensitive to temperature change than a metallic resistance (as used, for example, in the normal resistance thermometer).

Thermistors are one of the few semiconducting components manufactured which are not made from single crystals. Although a slice of intrinsic germanium has a useful resistance against temperature characteristic (Figure 1.13) much research has enabled reliable and cheap components to be made from a mixture of semiconducting oxides, commonly Fe_3O_4 and $MgCr_2O_4$.

The resistance of a commercial thermistor is normally specified to be within ± 20 per cent of a stated value at a given temperature. On occasions the resistance against temperature characteristic of a component may not be accurately reproducible.

A plastic binder added in manufacture to the carefully prepared oxide mixture enables the material to be extruded into rods, pressed into discs or formed into beads between two platinum wires. The component so formed is then sintered at high temperature which causes the material to shrink on to the wires and make good electrical contact. Miniature bead thermistors are often glass-encapsulated or enamelled for protection. A number of commercially available thermistors is shown in Figure 5.26 together with the conventional circuit symbol for a negative temperature coefficient device.

The relation between the resistance R and temperature T for a thermistor (Figure 5.27(a)) can be expressed in the form

$$R_T = A \exp(B/T) \qquad (5.5)$$

where R_T is the resistance at T K and A and B are constants for a particular component, A being in ohm and B in degree Kelvin. Hence

$$\frac{dR_T}{dT} = A \frac{d}{dT}\left[\exp\left(\frac{B}{T}\right)\right]$$

$$= -A \left[\frac{B}{T^2}\right] \exp\left[\frac{B}{T}\right]$$

$$= \frac{-AB \exp(B/T)}{T^2} = \frac{-BR_T}{T^2}$$

Hence

$$\alpha = \frac{1}{R_T}\left(\frac{dR_T}{dT}\right) = \frac{-B}{T^2} \qquad (5.6)$$

The constant B depends on the material. As α is specified as per degree Kelvin ($1/T$), it follows that B must be in degree Kelvin. Usual values of B are between 2000 and 6000 K.

Figure 5.27(b) shows the voltage against current characteristic of a thermistor being heated by the current passing through it.

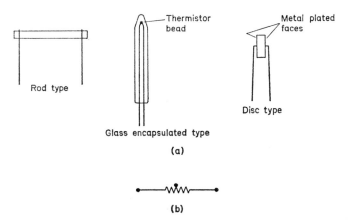

Figure 5.26. (a) Thermistors and (b) circuit symbol

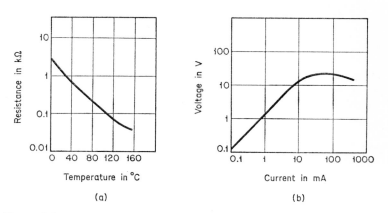

Figure 5.27. (a) Resistance against temperature characteristic of a thermistor and (b) voltage against current characteristic of a thermistor

Example 5.16

A negative temperature coefficient bead thermistor located in the end of a glass rod has a resistance of 2000 Ω *at* 300 K *and a value of B in equation* (5.5) *equal to* 2600 K. *Calculate its resistance at a temperature of* 500 K.

For a negative temperture coefficient thermistor, the resistance temperature relationship is given by equation (5.5):

$$R_T = A \exp(B/T)$$

At $T = 300$ K, substituting $R = 2000$ Ω and $B = 2600$ K gives

$$2000 = A \exp(2600/300) = A \exp(8.67)$$

At $T = 500$ K, the resistance is R_{500} given by

$$R_{500} = A \exp(2600/500) = A \exp(5.2)$$

Hence

$$\frac{R_{500}}{2000} = \frac{\exp(5.2)}{\exp(8.67)} = \exp(5.2 - 8.67)$$

$$= \exp(-3.47) = 0.0311$$

$$R_{500} = 2000 \times 0.0311 = 62 \text{ Ω}$$

5.17 Some Applications of Thermistors

The most obvious application for a thermistor is to the measurement of temperature. Any one of the three circuits shown in Figure 5.28

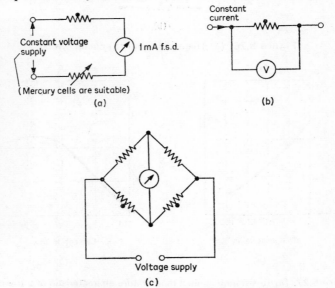

Figure 5.28. Thermistors in temperature measurement

can be used; the current through the thermistor should be kept as small as possible. The bridge type circuit (Figure 5.28(c)) is very sensitive and also compensates for changes in the ambient temperature.

The following advantages result from the use of thermistors for temperature measurement:

(a) The thermistor has a useful temperature range: −70°C to 300°C.
(b) It has high sensitivity: −6 per cent change in resistance per degree Celsius change in temperature at room temperature.
(c) Because of the high resistance of the thermistor, the length of the connecting leads and their change of resistance with temperature is generally of no consequence.
(d) A small thermistor bead has a very small heat capacity as a temperature sensing device. It is consequently able to respond to rapid changes of temperature.
(e) The element is more robust than the thermocouple and the platinum resistance thermometer.

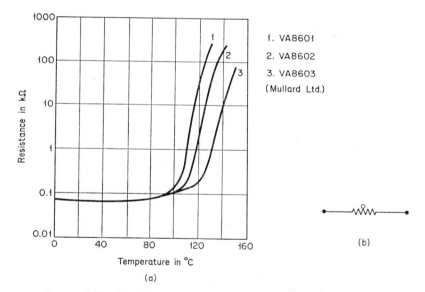

Figure 5.29. (a) Characterisics of positive temperature coefficient thermistors and (b) circuit diagram symbol for a positive coefficient thermistor

A thermistor thermometer has to be calibrated; in a simple experiment this is done against a mercury-in-glass thermometer.

Thermistors are frequently used in conjunction with relays. For example, a sensitive fire alarm circuit has a thermistor in series with the energizing coil of the relay. When the thermistor is at room temperature its resistance is large enough to restrict the current to below the value at which the relay is energized. Should the temperature of the surroundings rise, the resistance of the thermistor will fall, the relay becomes energized and the current to the fire alarm bell (which is much larger than the current in the relay energizing coil) is switched on.

Thermistors with *positive* temperature coefficients of resistance are available but the range of resistance values is severely limited. They are made of barium titanate.

Below a certain critical temperature, the resistance of such a thermistor is almost constant. Once the critical temperature is exceeded, the resistance rises rapidly with temperature increase. By controlling the process of manufacture of this type of thermistor, the critical temperature can be arranged to be for example, 100°C, 115°C or 130°C (Figure 5.29(a)).

A positive temperature coefficient thermistor has a circuit diagram symbol similar to that of the negative coefficient device except that an open circle on the resistor symbol is used in place of the black dot (Figure 5.29(b)).

Exercise 5

1. Describe the structure and the emitter characteristics of a unijunction transistor (UJT). Draw a circuit diagram of a UJT relaxation oscillator, and sketch the voltage waveforms across the capacitor and across the resistor in series with base 1.
2. Define the terms *peak point voltage, intrinsic stand-off ratio* and *equivalent diode voltage* with reference to a unijunction transistor (UJT). Outline the principles of a frequency divider network which utilizes a bipolar transistor and a UJT.
3. Describe the structure and the properties of a silicon controlled rectifier. Explain what is meant by a.c. phase control and outline the advantages afforded by silicon controlled rectifiers in power control circuits.
4. Draw the circuit diagram of a bistable circuit based on the use of silicon controlled rectifiers and explain the action of this circuit.
5. 'The principle of operation of a field effect transistor is quite different from that of a bipolar junction transistor'.
   Provide evidence in support of this statement.
6. With the aid of an appropriate circuit diagram explain how the drain characteristics of an n-channel field effect transistor are determined. Sketch the drain characteristics and comment on the prepinch-off and pinched-off regions.

Other Semiconductor Devices

7. Outline the working principle of a silicon surface junction particle detector. Why is the energy resolution obtainable with a solid state detector superior to that of earlier detectors?
8. The reverse current through a junction diode increases considerably when visible light falls on the junction. Why does this occur?
(A.E.B., part)
9. The energy gap (E_g) in germanium is 0.75 eV. Calculate the maximum wavelength radiation which is capable of creating an electron-hole pair in germanium. (The Planck constant, $h = 6.62 \times 10^{-34}$ J s; 1 eV = 1.6×10^{-19} J; velocity of light in free space, $c = 3 \times 10^8$ m s^{-1}).
10. (a) Why are transistors mounted within air-tight and light-tight containers?
    (b) Why is silicon usually preferred to germanium as a transistor material?
11. Sketch a curve showing the variation of resistance with temperature for a negative temperature coefficient thermistor. What is the form of the equation that approximates to this curve?
12. For the construction of a device for sensing temperature changes, compare and contrast the use of (a) a metal and (b) a thermistor material where, in both cases, the variation of resistivity with temperature is determined.
13. What advantages and disadvantages result from the use of thermistors for temperature measurement?

Answers

Exercises 1
3. 3.5×10^7 m s^{-1}
5. 7.5×10^{-2} m
15. 3.8×10^{20} m^{-3}
17. 3.2×10^{21} m^{-3}

4. 0.454 km s^{-1}; 8.5×10^{-5}
10. 0.282 nm
16. 2.5×10^{19} m^{-3}

Exercises 3
6. 35.3 mA
9. 30 µF

7. 1 : 0.81

Exercises 4
18. 2.0 V; 2.0 mW

Exercises 5
9. 1.65×10^{-6} m

Index

ACCEPTOR impurity, 30
Alternating current, 71–72
 average value of rectified, 72
 phase control, 145–148
 root-mean-square value, 72
Ampere, 1
Amplifiers,
 common-emitter connection of p–n–p transistor, 109–115
 difference, 129
 equivalent circuit of a common-emitter amplifier, 114
 feedback, 122–124
 flat response, 116
 open-loop gain, 124
 two-stage common-emitter transistor, 121–122
Ångström unit, 4
Astable circuit, 133
Atomic number, 11, 15, 19
Atom, nuclear model, 11
 shell structure, 16–20
Avogadro constant, 8

BARDEEN, J., 155
Bel, 115
Bipolar transistor, 154, see transistor, bipolar junction
Bohr, N., 33
Boltzmann constant, 8, 10
Brattain, W. H., 155

CADMIUM sulphide, 25, 168
 exposure meter, 169
Capacitor filter, 80–84
Carrier mobility, 31
Choke filter, 84–85
Conduction band, 38
Conductivity, 22

Constant-current source, 124–126
Covalent bonding, 27
Crystal growing, 56
 Czochralski method, 56
Crystals, 3

DE BROGLIE, L., 1
Decibel, 115–116
Depletion region, 70, 155
Dirac, P., 10
Donor impurity, 29
Doping, 29

ELECTRIC field strength, 5
Electron, 1
 arrangement in atoms, 15–21
 charge, 1
 free, 1
 mass, 1
 motion in electric field, 5
 motion in magnetic field, 14
 number per coulomb, 1
 specific charge, 7
Electron-hole pairs, 28
Electron-volt, 10
Energy gap, 37
 germanium, 38–39
 insulator, 37
 silicon, 38
Energy levels in isolated atoms, 33
 bands in semiconductors, 38–39
 bands in solids, 33–39
Extrinsic conductivity, 28

FEEDBACK, 122–124
 factor, 123
 negative, 122
 positive or regenerative, 122
Fermi-Dirac distribution, 24

Fermi, E., 10
Floating zone process, 55
Fourier analysis, 80

GALLIUM arsenide, 25
Germanium, 20, 25–26, 55
  crystal lattice, 27
  extrinsic, 40
  intrinsic, 40
  refining, 53, 54

HALL coefficient, 46
Hall effect, 26, 44
  identification of majority carriers by, 48
Hole, 27
Hydrogen atom, 12, 33–34

IDENTIFICATION of n or p type material, 50
  by cooled probe, 50
  by Hall effect, 48
Impurity conductivity, *see* extrinsic conductivity
Indium antimonide, 25, 168
Insulators, 1, 36
Integrated circuit, 61
Intermetallic compounds, 25
Interstitial impurity, 4
Intrinsic conductivity, 28
Ionization, 11, 30
  potential, 13

KILOMOLE, 8
Kinetic theory of gases, 2, 7–11

LEAD sulphide, 168
Leakage current, 116–118
Load line, 111

MAGNETORESISTANCE, 26
Majority carriers, 30
Matter, states of, 2–3
  structure, 1–11
Maxwell–Boltzmann distribution, 24
Mean free path, 9–11
  electron, 21–22
Metallic conduction, 1, 21–24, 33
  free electron gas theory, 2, 21–24

Minority carriers, 30
Multivibrator, 133–136
  long period, 136

OHMIC contact, 26
Oscillators, transistor, 130–136
  crystal-controlled, 131
  phase-shift, 132–133
  RC sinusoidal, 130–131
  relaxation, 142

PENTAVALENT doping elements, 29
Periodic table, 15, 18–19
Photoconductive cell, 168–169
Photoconductive effect, 26
Photodiode, 30, 166, 167
Photoelectric effects in semiconductors, 165–169
Photon, 11
Phototransistor, 30, 166, 167–168
  light meter, 168
Photovoltaic effect, 26, 166
Planck constant, 12
p–n junction, 53
  current, 68–69
  diode, 66, 69–71
  forming, 57–60
Point contact diode, 71
Positive hole, 28
  motion in electric field, 28
Positive ion, 11
Power supply, 62–64
  stabilized, 126–129
Proton, 16

QUANTUM numbers, 16–17

RECTIFICATION, 62
  full-wave, 63, 74
  half-wave, 63, 72
Rectifier, 62
  moving-coil instrument, 77
Reference-voltage diode, *see* Zener diode
Resistivity, 22
Ripple voltage, 82
Rutherford, Lord, 11

SEMICONDUCTOR devices, manufacture of, 53–61

Index

Semiconductor diodes, 62–71
  characteristics, 66–69
  electrical behaviour, 69–71
  forward biased, 66
  reverse biased, 66
  reverse saturation current, 69
Semiconductor particle detectors, 163–165
Semiconductors, 1, 24
  compound, 25
  doping, 29
  effect of impurity atoms, 26
  effect of surface state, 26
  extrinsic, 28, 38–43
  intrinsic, 28, 38–40
  near-intrinsic, 30, 32
  n-type, 29
  photoelectric effects, 165–169
  p-type, 30
Shockley, W., 155, 156
Silicon, 20, 25
  refining, 53–54
Silicon controlled rectifier, 65, 145–154
  a.c. phase control, 145–148
  bistable circuit, 149–150
  conduction angle, 146
  demonstration of phase control, 151–152
  gate, 145
  holding current, 145
  lamp dimming circuit, 152
Silicon diode, 64, 65, 67
Silicon planar diode, 60
  passivated, 61
Sinusoidal waveform, 63
Smoothing circuit, 79
  capacitor, 80–84
  choke, 84–85
Solar battery, 166
Solid state, 1, 33
Sommerfeld, A., 10
Stabilized supply, variable, 91
Substitutional impurity, 4

TEMPERATURE coefficient of resistance, 169
  insulators, 37
  metals, 15
  semiconductors, 26, 39

Temperature—(contd.)
  thermistors, 169–170
Thermal runaway, 116–118
Thermistors, 169–174
  applications, 172
  circuit symbols, 171, 173
  resistance, 169–171
  resistance against temperature characteristic, 171
  voltage against current characteristic, 171
Transistor, bipolar junction, 97
  base, 97
  biasing, 118–121
  collector, 97
  collector-base junction, 97
  common-base connection, 99, 104
  common-emitter connection, 99
  comparative features in CB and CE connection, 107
  current gain, *see* forward current transfer ratio
  emitter, 97
  emitter-base junction, 97
  forward current transfer ratio, 101, 106, 107
  hybrid parameters, 108
  incremental resistance, 100
  input characteristics, 104
  input resistance, 100, 104
  leakage current, 116–118
  n–p–n type, 57, 65, 97, 98
  outlines (TO), 64, 66
  output characteristics, 101, 102, 105
  output resistance, 101, 105
  p–n–p type, 57, 65, 97
  power, 59
  regulator, 127
  relationship between h_{FB} and h_{FE}, 107
  slope resistance, 100
  tester, 93, 136–138
  thermal runaway, 116–118
  transfer characteristics, 102–103
Transistor, field effect, 154–160
  channel, 159
  circuit symbols, 159
  drain, 157, 158

Transistor—(contd.)
  drain characteristics, 160
  gate, 157, 159
  model, 157
  n-channel, 154
  p-channel, 154
  pinch-off effect, 158
  polarity conventions, 159
  special features, 163
  structure and behaviour, 156
  source, 157, 158
  transfer characteristics, 161
  voltmeter, 163
Transistor, unijunction, 140–145
  frequency divider, 144
  inter-base resistance, 140
  inter-base voltage, 140
  intrinsic stand-off ratio, 141
  relaxation oscillator, 142
  staircase generator, 143
Transverse galvomagnetic effect, *see* Hall effect
Trivalent doping elements, 29
2–6 compounds, 25

UNIPOLAR device, 154

VACANCY, 27
Valence band, 38
  electrons, 17
Varactor diode, 71
Virtual probe, 47
Voltage doubling circuit, 77–78
Voltage quadrupler circuit, 79
Voltage regulator diode, *see* Zener diode
Voltage stabilization, 64, 126–129

WIEDEMANN–Franz law, 23

ZENER diodes, 64, 84, 86–94
  in parallel, 91
  in series, 90
  limiter, 92
  protection of voltmeter, 91
  shunting of meter, 92
  voltage regulating circuits, 88–90
Zinc sulphide, 25
Zone refining, 30, 54